Paul Stull

Construction Surveying & Layout

 Building News

ANAHEIM - NEW ENGLAND - WASHINGTON, DC - LOS ANGELES

BNi. Building News

EDITOR-IN-CHIEF
William D. Mahoney, P.E.

TECHNICAL SERVICES
Rod P. Yabut

COVER DESIGN
Robert O. Wright

LOS ANGELES
10801 National Blvd, Suite 100
Los Angeles, CA 90064

NEW ENGLAND
PO BOX 14527
East Providence, RI 02914

ANAHEIM
1612 S. Clementine St.
Anaheim, CA 92802

WASHINGTON, D.C.
502 Maple Ave. West
Vienna, VA 22180

1-800-873-6397

ISBN 1-55701-363-2

Contents

1 Introduction to Land Surveying 9

    The History of the U.S. Land Surveys 10

    Land Deeds 14

2 The Building Site 19

    Type of Development 20

    Choosing a Site 20

    The Next Step 24

3 The Survey **25**

    Surveying Instruments 26

    A Practice Survey..27

    Closing 30

    Economizing in the Field 34

4 Using a Transit and Tape **43**

    Reference Lines...44

    Laying Out Roads 51

    Laying Out a Line 51

    Setting Parallel Lines 58

    Planning a Bridge 61

5 Applied Geometry Using a Transit **65**

    Using an Established Point 66

    Finding a Perpendicular Line..66

    Establishing Parallel Lines..67

    Dividing a Line Into Equal Parts..68

    Drawing a Tangent Circle 69

    Laying Out Angles 71

    Drawing a Circle Through Three Points 73

Drawing a Hexagon 74

Ellipses 75

Pentagons 77

The Involute of a Circle 79

6 Stadia Surveying **83**

The Tools 83

Geometry You will Need to Know 85

Practical Considerations...92

7 Topographic Surveys **95**

Illustrating a Depression 96

Making a Topographic Survey..97

8 Leveling...**105**

Leveling Instruments..105

9 Basic Leveling Surveys **109**

Finding Elevations 110

Types of Leveling 112

Grade Stakes 113

Potential for Error 115

10 Complex Leveling...**117**

Evaluating a Site 118

Planning the Streets...118

Grading 123

Necessary Curves 128

11 Surveying Notes **145**

12 Special Problems **147**

Finding Missing Measurements 148

Subdividing Land 162

13 Mapping the Site **167**

Mapping Procedure 168

General Points 169

Before Construction Begins 170

14 Plotting Angles **171**

    Using a Protractor 172

    Using Trigonometry..172

15 True North, Latitude, and Longitude **177**

    Locating True North..178

    Locating by Latitude and Longitude 182

Appendix...**185**

    A. The Transit 187

    B. Geometry for Construction..193

    C. Trigonometry for Construction 225

    D. More Practical Examples 231

Index..**241**

Preface

Your community library probably has several books on surveying. Many have been written. But nearly all are technical and detailed, offering far more theory and much greater depth than builders and contractors need. That's too bad, because most builders and many tradesmen need a working knowledge of basic survey principles.

On every job someone has to find or verify the location of lot corners, align foundation, walls and floors, lay out perpendicular and parallel lines, mark angles, set grade or calculate cut and fill quantities. Professional surveyors can do this work, of course. But you do not need to hire a professional to check a boundary line or lay out a driveway. After reading this manual, I think you will agree that most of the survey and layout work on a construction site is relatively easy. Anyone with the desire and time available can master the skills required.

This manual should meet your needs precisely if you want to learn construction surveying and layout, but do not have the patience to wade through a detailed, theoretical surveying text. I will emphasize the practical rather than the theoretical and focus on the type of survey and layout problems you are likely to face on your next job.

If knowing how to make a simple survey will help in your construction work, you are reading the right book. I will cover all the practical surveying and layout you are likely to need for any construction project. If you are concerned that your math skills may not meet the challenge presented by the more complex survey problems, do not worry. I've simplified the trigonometry, geometry and mathematics throughout this volume. If you come to a symbol or calculation you do not understand, a little study of the appendix should clarify the point. Appendix B summarizes all the geometry you need to know and Appendix C explains the essentials of trigonometry.

Before we get into the first chapter, let me emphasize how important good survey practice is in construction. We've all heard sad stories about homes, apartments or garages that have been built on the wrong lot or straddling a property line. Fortunately, that's not a common error. But it is nearly always a very expensive mistake. Much more usual are foundations that aren't level, slab corners that aren't square, circles that are more like ellipses and ellipses that aren't like anything at all. These all begin with survey mistakes. But that's just the beginning. Anything that's not level, square and true in construction tends to affect every part of the job that follows. You begin with a foundation wall that isn't square at a corner and end up with roof sheathing panels that do not fit right.

The surveyor on the job has the first chance to make a mistake. If he or she does it right, the first error has to be made by someone else. Survey and layout that are done with care and professionalism promote craftsmanship throughout the project. What's the best way to be sure the survey and layout are done right? It is to do it yourself, or at least check it yourself. Maybe that's why some of the best, most successful contractors and builders I know have taken the time and trouble to learn construction surveying and layout.

Having said that, it is time to get down to business. We'll begin Chapter 1 by introducing land survey terms and concepts. By the time you have worked your way through this manual to the end of Chapter 15, I think you will agree that construction surveying doesn't have to be difficult. There's nothing technical or theoretical here, but there's a lot to remember. Between Chapter 1 and Chapter 15 I will explain it all, easily remembered.

CHAPTER 1

Introduction to Land Surveying

Land surveying allows a surveyor to precisely determine the area of any part of the earth's surface, the lengths and bearings (direction) of all the boundary lines, the contours of features of the land, and to accurately describe all of this information on a map.

A knowledge of surveying is indispensable if you are a builder, contractor, or developer. Not only will you often need to hire surveyors, check their findings, and read survey data and maps, occasionally you may be both the builder and the surveyor on a project. There's certainly no better way to find out about every aspect of the site you are developing.

This book describes the mathematics behind surveying, the instruments used to take measurements, the method for surveying a site, and the procedure for drawing accurate maps. You will learn how to take complex measurements for roads, buildings, and bridges; how to establish grades for areas of cut and fill; and how to solve problems that may occur during a survey. The first thing you will find out is how surveys actually started in the United States.

The History of U.S. Land Surveys

Land surveys made by government authority follow a definite system provided by law. These are United States Land Surveys. They began on May 18, 1775, when the Continental Congress called for a method of measuring and disposing of some lands claimed by the state of Virginia, which at that time included areas of West Virginia, Kentucky, and Ohio.

■ The Rectangular System

Accordingly, the *Manual of Instructions for the Survey of the Public Lands of the United States* was developed. It describes the conditions of the *rectangular system of surveys* that is required for U.S. land surveys.

Longitude and Latitude. The rectangular system uses the longitude and latitude lines of the earth as its base. The longitude lines, called *meridians*, run north and south from pole to pole and at 90° to the equator.

Latitude lines, called *parallels of latitude*, run east and west around the earth parallel to the equator. The equator is 0°. Parallels of latitude are 90° north of the equator (north latitude) and 90° south of the equator (south latitude). Longitude and latitude are stated in degrees, minutes, and seconds.

Initial Points. Under this system the initial point of a survey is referenced by longitude and latitude by astronomical methods, and marked accurately, called *monumenting*. This is done so that it is easy to identify exactly where the survey was started and, thus, exactly which piece of land is being described.

Initial points were established throughout the public domain by monumented meridian lines and baselines. Each meridian is identified by a name and number. These are listed in the manual of instructions.

All of the lines in the rectangular system are tied into or extended from these principal meridians and the baselines. They are identified as guide meridians and standard parallels, township exteriors, section lines, and meander lines (established by the water line of a lake or stream). In recent years, county officials have required that surveys be referenced to the legal corners of townships, and bearings and distances given in the deed (legal description) such that the surveyed area be satisfactorily closed by one of several methods (which will be explained later).

Land Divisions. The rectangular system was first used to survey Ohio. It began at the point where the Ohio River leaves Pennsylvania.

The public lands are divided into counties. Counties are divided into townships. Townships (6 miles square) are divided into 36 sections (1 mile square). Sections are divided further into quarter sections.

And, quarter sections are divided into quarter-quarter sections (see Figure 1-1). As always, there are a few exceptions to the rules. For instance, in Louisiana the word *parish* is used instead of *county*. And, sometimes townships aren't divided into exactly 36 sections.

A county contains townships, and a north-south row of townships constitute a *range of townships*. An east-west row of townships constitute a *tier of townships*. Figure 1-2 shows the arrangement of the 36 sections in a township.

Figure 1-3 illustrates the political subdivision of land. The township boundaries running north and south, as *ab*, are intended to be true meridians and are called *meridional boundaries*. A series of townships in a north-and-south row constitute a *range* of townships. Therefore, the meridional boundaries are commonly known as *range lines*.

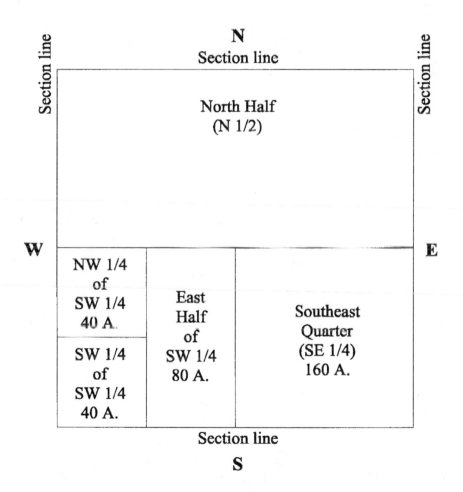

Figure 1-1 *A Typical Section*

Townships are numbered as ranges in both east and west directions. Here they run west from the principal meridian, hence the letter W. Townships also are numbered as tiers both north and south, beginning with number one at the baseline.

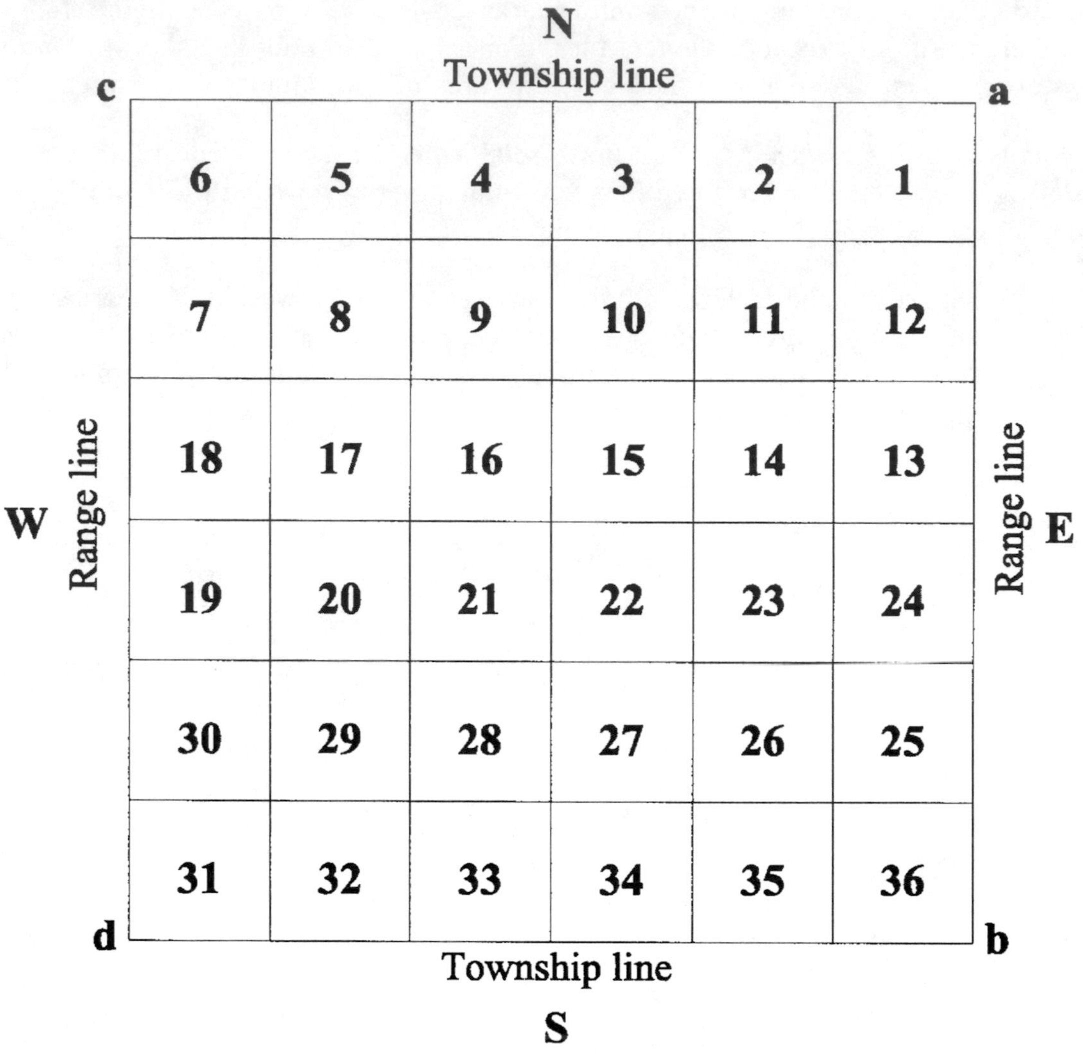

N

Township line

c						a
6	5	4	3	2	1	
7	8	9	10	11	12	
18	17	16	15	14	13	
19	20	21	22	23	24	
30	29	28	27	26	25	
31	32	33	34	35	36	
d						b

W Range line   Range line **E**

Township line

S

Figure 1-2 Township and Range Lines

You can see that Jackson Township in the northwest corner is in Range 19 West (R-19-W), and in Township (tier) 7 North (T-7-N).

Look north to south along the west county line. You will see an example of mixed section numbering shown along the Jackson and Liberty townships. In Jackson Township, section 19 has a section 20 on either side of it. Section 30 has a section 31 on either side of it. This seemingly odd-numbered section is due to a land trade made with Pike County many years ago. You will find this jumbled numbering in many places. There

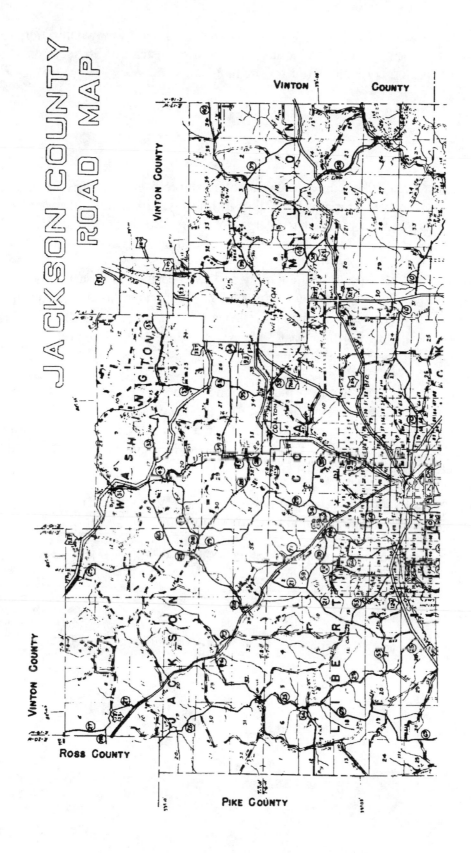

Figure 1-3

are localities in Ohio where the range lines run east and west and the township tiers run north and south. And, some townships there contain 25 sections instead of 36 sections. So, when you are researching area records, look out for the unusual.

Land Deeds

This rectangular system of surveying with official requirements was created because surveys are used as the basis for the land descriptions used in deeds. A *deed* is a legal document describing a certain piece of land that you own. So, if everyone used a different method of surveying for describing that land, no one would be sure who owned what.

Deeds describe the initial point of the survey and use bearings to describe the boundaries of the land in question. Figure 1-4 shows how a survey plane is divided into four quadrants, each 90°. The bearing is then referenced for every case, from the north-south line. For instance, bearing N30°E runs north by east 30° from the north-south line and bearing S30°W runs south by west 30° from the north-south line.

Suppose that you are about to buy a piece of land in the northwest area of the south half of the township shown in Figure 1-3, which is in Range 19 West of Jackson Township. You and the owner have visited or driven around the area you plan to buy. You then hire a surveyor to describe the boundaries of that piece of land. Your deed to the land will be written from the description made by the surveyor.

Since this parcel of land is in the northwest quarter (NW 1/4) of the southwest quarter (SW 1/4), the introductory paragraph of your land deed will read:

Being a parcel of land in the NW 1/4 of the SW 1/4 of Jackson Township, Jackson County, Ohio, R-19-W, T-8-N.

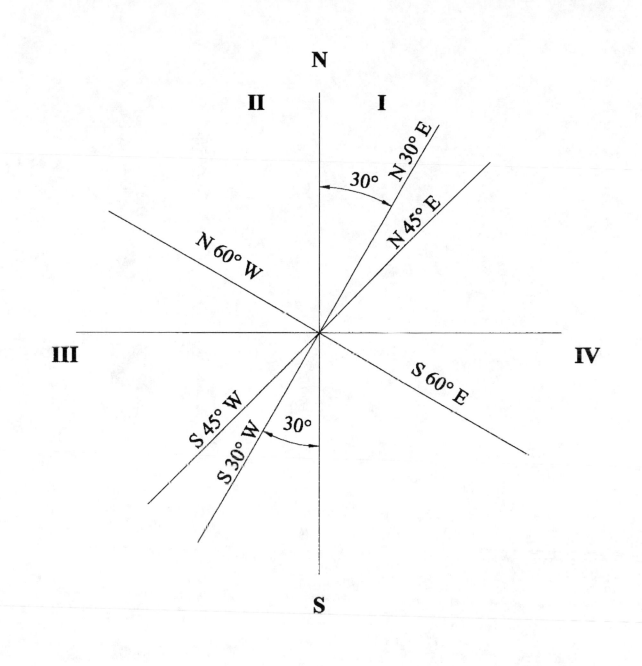

Figure 1-4 Survey Plane

You can see the T-8-N (Township Eight north) on the south line of the township. This is an accurate legal description, written from a correct survey, that identifies this parcel of land from any other parcel anywhere on earth (See Figure 1-1).

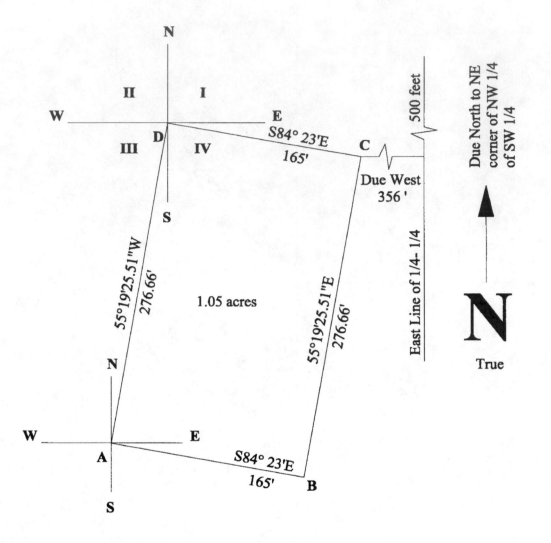

Figure 1-5 Survey sketch showing bearings

To make it even better identified, the survey is tied to an initial, permanent point (described previously). In this case, the legal corner (any corner of a political subdivision) provided by the northwest corner of the NW 1/4 of the SW 1/4 of Jackson Township, is a perfect reference point for the survey of your land. Therefore, the legal description following the introductory paragraph in your deed reads as follows.

Beginning, for reference, on a stone at the NE corner of the NW 1/4 of the SW 1/4 of Jackson Township,

Thence 500 feet due south along the east line of said 1/4-1/4 to a point in said line;

thence 356 feet due west to an existing fence corner, said corner being the place of beginning for this survey.

thence N84°23'W along a fence line a distance of 165 feet to a point in said fence line;

thence S5°19°25.51"W a distance of 276.66 feet to a point in an open field;

thence S84'23'E a distance of 165 feet to a point in an open field;

thence N5°19'25.51"E a distance of 276.66 feet to a point, said point being the fence corner at the place of beginning for this survey, said survey containing 1.05 acres, more or less.

Notice that in descriptions of a survey bearing, the words north, south, east, and west, are not used alone. For instance, the bearing on the east line of the 1/4-1/4 (Figure 1-5), is described as being "due north." This is the back bearing for the bearing given as "due south" in the deed description. Always use the word "due" with the cardinal direction.

Now you know how surveys began and how they are used to write deeds. The next step is to find out how to do a survey using the instruments and methods described in the following chapters.

CHAPTER 2

The Building Site

This chapter explains the various factors that go into the choice of a building site. In order to create a high-quality development that is profitable for you and pleasant for future occupants to live and work in, you need a great deal of detailed information. But by asking the right questions and finding people with answers, you can proceed from research to construction to sales without a hitch. So, before you start surveying, carefully choose your site.

Type of Development

Before you can choose a site, you need to evaluate the type of building you are planning. For instance, are you planning to build residential homes? If so, the following are just a few of the questions you will need to ask.

1. What will be the income bracket of the owners?

2. Will there be young couples with growing families?

3. Will there be older couples?

4. Will there be single, senior citizens?

5. Will it be low-income, subsidy housing?

6. Will there be retirement units?

These factors will determine the size, location, and number of building sites you will need. They'll also be important in choosing the design and materials for the structures and the layout of the entire development. Therefore, before you start choosing sites, carefully consider all related factors.

Choosing a Site

■ Visits to Agencies

After you have decided on the type of development you will build, you can start shopping for suitable building sites. A good place to start is at the county engineer's office. There you will find many maps of the area and land records. You will be able to find maps that include the names of the property owners of the land that interests you. And, you can start answering more questions, such as these.

1. Is the area city, suburb, farmland, or wasteland?

2. Are there intermingled houses and small businesses?

3. Is the land level or hilly? (If the area is not level, surveying will be more difficult and building costs will be higher because earth moving will be needed.)

4. Are there highways located nearby?

5. Are there streams or other bodies of water? (Could flooding be a problem? Will bridges be needed?)

Be sure to talk with people in the engineer's office. Many of them have worked there for years and know the area. Discuss it with them in as much detail as possible.

When you have chosen an area that really interests you, look a little further. Check with the county recorder to see a copy of the property deed. Here careful reading can reveal any inconsistencies. Some deeds will tell you only that the property is out there someplace. For example, ". . . all that land west of John Doe, north of Edward Doe, east of James Doe, and south of Charles Doe." This description might have been written long ago and carried down through the years to the present owner. If so, a true description of the land is anyone's guess. Examine the deed for possible restrictions, or any encumbrance of any nature. And then have an attorney make an abstract of the property before you get further involved.

Talk with people who work in the recorder's office too. They have many contacts with the legal and business communities, so they may know of other developments planned near or in the area.

Finally, the county auditor can supply you with property value trends. Knowing the property values will help you to determine whether the neighborhood is suitable for the type of development you are planning. Ask about tax rates and other possible assessments while you are there.

■ Visiting the Site

Next, visit the site. Drive around it. Drive over every area. Inspect the neighborhood. Is it clean, well kept? Talk with adjacent property owners and ask for their views on the advantages and disadvantages of living there. Ask adjacent owners if they agree with the existing property lines. Ask if there has ever been a dispute over the property lines. If so, was it satisfactorily settled? You need to know these answers because your prospective buyers will probably ask these questions. If you are happy with what you see and hear at the site, you will want to proceed.

■ The Preliminary Survey

The next step is to do a preliminary survey to determine if the existing property lines correspond with the deed description. First, you should clear the survey with the present owner.

■ Further Considerations

Utilities. Check on the availability of utilities and public services. Determine what it will cost to develop these if they aren't available. If the costs will be prorated per unit, the market price might be too high to be competitive.

Find out if the electric lines are overhead or underground. Some companies are installing lines underground. Does the builder (you) pay a portion of the cost? Remember that increasingly, telephone lines are going underground. You may need to provide utility space not only for the lines but for distribution structures and test points as well.

Water pressure as well as water supply are important to you. Many communities have sufficient water but have overextended and outdated equipment which mean a resultant drop in pressure. Can the utility company guarantee acceptable pressure? If not, the additional demand from your development could noticeably weaken the pressure for everyone.

Additionally, even if the water pressure is fine for domestic use, find out if it will maintain pressure for fire protection. The demand for a fire could create an unstable water supply. Check with the National Board of Fire Underwriters to find out what the established fire flow is by population size.

Sewerage Systems. Check to see that a central sewage-treatment plant is available and that sewerage from your development can be routed to it. While you are planning costs, remember that it is almost a certainty that initially you as the builder will have to pay for sanitary and storm sewerage costs. Later, a proportionate share will be allocated to each house, to be paid by the owners.

If a sewerage-treatment plant isn't available, the alternative is on-site sewage facilities. Depending on local health laws, on-site treatment can be of three types: aeration, lagoon, or septic tank and leaching field. The last two choices have been the subject of much controversy and experimentation. Talk with the Environmental Protection Agency. This agency has completed considerable research of on-site treatment in recent years and can offer you excellent technological.

Before you choose any sewage system, be sure to check on local laws and requirements. Leave no unanswered questions and be sure to check with health officials because they will inspect both the site and the plan.

Heating. Home heating is of great concern to every home owner in a cold climate. Therefore, if the site is in an area with a medium or long winter, give the type of heating careful study. There are five commercial types of heating: electric, oil, coal, natural gas, and synthetic gas. Less used options are solar and wind-power heating.

Supply all of the utility companies with a layout of the lots so they can plan a route for utility lines. Include each layout on the site plans.

Waste Disposal. Find out if solid-waste disposal collection is available. If not, it is probably left to each home owner. This is an important item.

If municipal collection is not available, investigate possible private collection services. To protect the community strong regulations to govern waste disposal should be formulated by either local or state government.

Mail, Police, and Fire. Find out whether mail delivery is by foot carrier or rural route. Keep in mind that a mail box must be provided either at the curb or at the house.

Police and fire protection are generally provided by either the city or the county. Determine which type of protection you may need during construction.

Road Repair. Inspect the extent of road repair on the approach roads. Find out which agency is responsible for road repair and snow removal. Will the same agency take care of future roads on your site? Check with area residents to find out how well current roads are cared for and find out what can be done to repair any problem areas. The condition of the approach roads could be a factor in future sales.

Schools. Is there a local school system? What do the area residents think of it? Visit the school and the school board during a meeting. Find out all you can about the system; prospective home buyers will.

Easements. Find out whether any easements for utility right-of-way exist. It could be disastrous to your site plans if you find out too late.

Drainage. Run a drainage survey on a topographic map of the area. Also, be sure to check topographic conditions visually at the site to determine whether the map is correct. Previous excavation on the site could have altered drainage capacity to a considerable extent.

Does the area drain well during wet seasons? Can it be made to drain well without excessive cost? Flooding due to poor drainage or poor storm discharge pathways can damage property. A creek, ravine, or even a well-constructed ditch can flood the site if the discharge point is already flooded with storm water or if its flow is impeded by poor trunk drainage.

Other Questions. The following is a list of other questions you might need to consider for a thorough evaluation of the site.

Building related:

         ☑ How many construction permits are required?

         ☑ What will they cost?

         ☑ Is the site on a landfill?

         ☑ How is the site oriented in terms of the wind?

☑ How is the site located in terms of the sunlight?

☑ Are there any restrictions on building materials?

Questions home buyers will ask:

☑ What are zoning regulations regarding plants and animals?

☑ Are there high-voltage lines nearby?

☑ Are there high-pressure gas lines?

☑ What is the crime rate?

☑ Are there odor-producing agricultural facilities nearby?

☑ Is there a high-pressure deep well that receives hazardous waste?

☑ Is there a solid-waste landfill nearby?

☑ Is airport noise a problem?

☑ Is air or water pollution a problem?

☑ Are there medical facilities nearby?

☑ Is it a reasonable distance to stores, churches, and recreational facilities?

☑ Can public transportation be extended to the new area?

The Next Step

After you have asked all the questions and found all the answers, you will do a complete survey of the area as discussed in the next chapter.

CHAPTER 3

The Survey

A survey consists of two major tasks: taking measurements of the land and using those measurements to draw an accurate map of the land. Taking measurements is called *field work*, and the most important element of field work is accuracy.

In order to get such accuracy, you use special surveying instruments that measure distance and elevation. There are three main tools: the transit, the level, and the measuring tape.

Surveying Instruments

■ The Transit

The *transit* is a telescope mounted on a tripod so that it can be turned at right angles over a horizontal east-west axis. The transit is used mainly to measure angles and distances by stadia. It is also useful for The Survey measuring short-distance elevations when the standard level is not available. The electronic transit measures both angles and distances. A compass on the transit is used to establish the north direction. Adjacent scales near the compass are used to measure angles. The scales are sometimes called "limbs."

■ Stadia

The transit also contains *stadia hairs* set horizontal in a transit, one above and one below the regular (centered) horizontal cross hair. With the stadia hairs and a specially marked *stadia rod*, you can find distances. Stadia rods are graduated in feet and tenths, which are represented by bold geometric figures designed for legibility at long distances. Hundredths must be interpolated. Some transits include a quarter-hair between the middle cross hair and the upper stadia hair for shots over 2000 feet. The ordinary leveling rod is satisfactory for short sights. The stadia method is especially useful in topographic surveys (see Chapter 6).

■ The Level

The *surveyor's level* consists of a spirit level and a telescope mounted on a tripod in a way that allows them to revolve around a vertical axis. The *spirit level* itself (or, simply, the *level*) is a device that you use to find a horizontal line (contour line) or plane by means of a bubble that moves to the center of a bowed, liquid-filled tube when you set it exactly horizontal. The level is used to find the difference between two elevations (see Chapter 8).

■ The Tape

The tape measure is a narrow strip (usually cloth, plastic, or steel), marked with units of measurement (feet or meters); it is usually used to find horizontal distances. Although it is a simple tool, the tape is indispensable for field measurements.

A Practice Survey

To get an overview of how a survey is done, assume that you are surveying a property in a typical township of 36 sections, as shown in Figure 1-1. The points *A, B, C,* and *D* in Figure 3-1 have been agreed on by the owner and the buyer as the corner points of the tract to be surveyed.

You find that the east line of the 1/4-1/4 runs due north and south. You start, for reference, at the northeast corner of the northwest quarter of the southwest quarter of the township. This is a legal corner. You move due south and then due west, measuring, taking notes, and sketching, as shown in Figure 3-1, and arrive at point C, the beginning point of this survey.

You continue from point to point (*C* to *D* to *A* to *B* to *C*) to establish the courses *AB, BC, CD, DA,* of the survey. Note that point *A* is the westernmost point and is used to begin the computations (see Figure 3-2). The bearings and distances are given in Chapter 1 in the section entitled "Land Deeds."

The lines at point *D*, marked *N-S, E-W*, and the letters *A, B, C, D*, marking the survey corners, do not appear on the map of a survey. They appear here only as reference points. They are used as shown in the left column in Figure 3-2 to identify the survey courses when computations are made.

∎ Running the Survey Lines

You set up the transit over point *C* and lock the scale reading zero to zero, then you backsight along the 356-foot due west line and lock the transit on this point. You flop the telescope (turn 180° in a vertical circle). With the transit still locked in place, loosen the upper-plate locking screw and turn horizontally to the right from the due west sight to point D. With the angle turned in a north direction, you read 5°37' from the scales. You are sighting into Quadrant II (See N60°W in Figure 1-4).

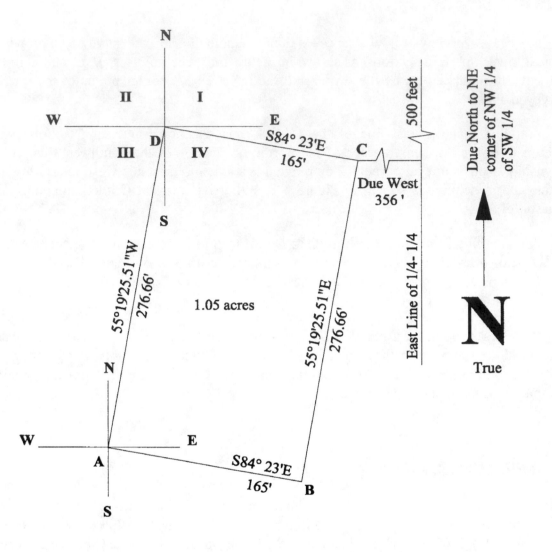

Figure 3-1 Survey sketch

The bearing is computed from the north direction. There are 90° in each quadrant. You subtract 5°37' from 90°. Thus, 90° translates to 89°60', because there are 60' in each degree. Then 89°60' minus 5°37' yields 84°23'. The complete bearing now is N84°23'W, because it lies in Quadrant II. You have obtained direction as well as degrees and minutes. The distance from point *C* to point *D* is 165 feet. You now move to point *D* where an entirely new set of conditions is encountered.

Course	Bearing	Distance Cosine	Sine	Latitudes North	South	Departures East	West	D.M. Distance	Double Areas North	South
AB	S 84-23 E	165	00		16.148	164.207		164.207 ⟋ 2 ⟍ 328.414 ⟋ + 25.669		- 2651.614
BC	N 5-19-25.51 E	276	66	275.466		25.669		354.083 ⟋ + 25.669 ⟍ 379.752 ⟋ - 164.207	+ 97,537.827	
CD	N 84-23 W	165	00	16.148			164.207	215.545 ⟋ - 164.207 ⟍ 51.338 ⟋ -25.669	+ 3,480.620	
DA	S 5-19-25.51 W	276	66		275.466		25.669	25.669		- 2,070.936
				291.61	291.61	189.88	189.88	(-) (÷2) (÷k)	+ 101,018.447 ⟍ 9,722,550 ⟍ 91,295.097 ⟍ 45,647.948 ⟍ 1.0479 Acres	- 9,722.550

Figure 3-2 Calculations for survey

Remember that your backsight is along the due west course. Because a due west course is 90° from due north, your bearing, 84°23', is from true north. You are now working in Quadrant III, in which all bearings are referenced to the south and west.

You backsight from point D to point C and then flop the telescope. Your telescope is again on the bearing 84°23'. You turn the angle 5°37' to the left, which puts the line of sight again along a due west course (84°23' plus 5°37' equals 90°). Once again, your scale reads zero to zero.

You unlock the upper scale and turn the telescope to sight on point A and read 84°40'34.49". Because there are 90° from due west to due south, you subtract. So, 89°59'60" minus 84°40'34.49" equals 5°19'25.51", which is the bearing of AD (or DA), S5°19'25.51'V. Distance is 276.66 feet.

A bearing is generally read to the nearest 20", because on many transits this is the lowest possible reading. The decimals 34.49 and 25.51 are found when you mathematically balance the survey.

You proceed to point A and set the transit over it and lock the scales zero to zero. Notice the north-south line through point A. You sight on point D, which places your line of sight along AD in Quadrant I.

Turn the angle 5°19'25.51" to the left to obtain a true north sighting. Read the angle as a bearing of N5°19'25.51"E, because you are now working in Quadrant I.

Turn a 90° angle to the right to obtain a true east direction, then sight on point *B*. You find you have turned an angle of 5°37', the same angle you turned when you sighted from *C* to *D*. Subtracting this from 90° yields 84°23', a line from *A* to *B* that is parallel to line *CD*. The distance is 165 feet. Because the lines *CD* and *AB* are on the same bearing and of equal length, *BC* and *AD* have identical bearings and distances.

■ Considerations in the Field

First, remember to keep the compass reading as true as possible. Never wear or carry, even in your pockets, anything metallic. Even a belt buckle is a problem. Clear all metal objects from the setup area. Your compass needle can easily be pulled away from magnetic north by local attraction from metal objects.

Second, when you prepare to survey a given tract of land, examine deeds or maps of adjacent properties and try to find a property-line bearing, either magnetic or true north, to use as a beginning bearing or azimuth (horizontal direction expressed as the angular distance between the direction of a fixed point and the direction of the object) for your survey.

Because there is a constantly shifting magnetic declination, a magnetic bearing or azimuth is likely to be different from day to day. If possible, use a true north bearing instead. But whichever you choose, be sure to make a note on the deed or survey map stating whether the bearings were taken from magnetic north or true north.

> **Note:** Because it is so much better, the rest of this book assumes you are using a true north meridian (see Chapter 15 to learn how to find one).

Closing

Closing a survey means checking the field measurements and finding the area. To close a survey, the north latitudes must equal the south latitudes and the east departures must equal the west departures.

Latitude is a distance north or south of a given point. Departure is a distance east or west of a given point. In Figure 3-3 the entire plot of land lies east of point A at the southwest corner. To calculate for closure, you should list the first course (bearing and distance) at the westernmost point of survey.

■ Finding Latitudes and Departures

The points *A*, *B*, and *C* in Figure 3-3 are identical with those in Figure 3-1. Point *A* is the westernmost corner of the plot of land as shown in Figure 3-1. The first course is *AB*; the second is *BC*; the third is *CD*; and the fourth is *DA*. List the courses in a counterclockwise direction.

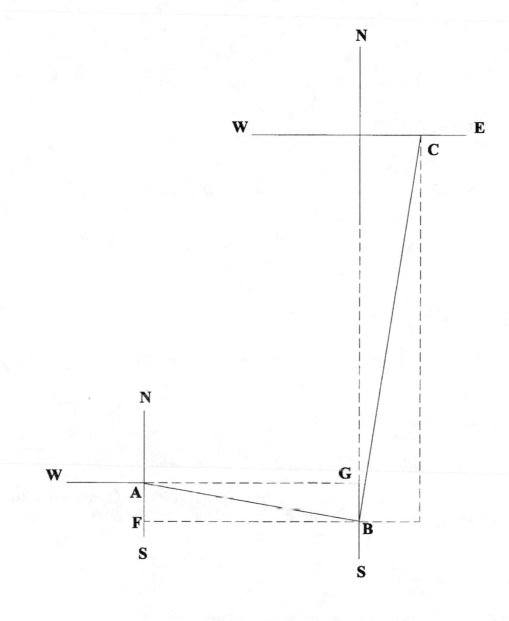

Figure 3-3 Westernmost corner of the example survey

Line *FB* though point *B* is parallel to line *AG*. This forms the rectangle *AFBG*. The angle *FAB* is 84°23'. Side *FA* equals side *GB*. Use trigonometry to find the latitude *FA*.

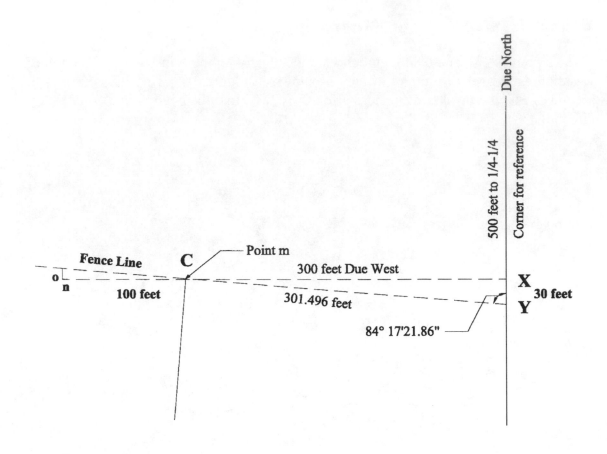

Figure 3-4 *Well-organized survey using few transit setups and distance measurements*

$$\text{Cos } 84°23' = \frac{\text{side adjacent}}{\text{hypotenuse}} = \frac{FA}{AB} = \frac{FA}{165}$$

$FA = .0978724 \times 165$ feet $= 16.14$ feet, the latitude for course *AB*.

The end of the course at point *B* is 16.14 feet south of point *A*. Course *AB* is 165 feet long, but point *B* is only east of point *A* the distance *AG* (or *FB*). The distance due east is the departure. Angle *FAB* is used to find the departure.

$$\text{Sin } FAB = \frac{\text{side opposite}}{\text{hypotenuse}} = \frac{FB}{165}$$

$FB = .995198 \times 165 = 164.207$ feet, the departure east of course *AB*.

■ Charting the Complete Survey

Figure 3-2 shows the calculations for the complete survey. The course and bearing are listed with the degree, minute, and second marks left off. Distances are listed with decimal digits under sine and the whole footage under cosine. Cosine, listed first, corresponds with north or south, listed first under latitudes. Sine corresponds with east or west, listed first under departures.

For course *AB* the measured distance, 165 feet, is multiplied by the cosine of the measured angle 84°23'. Since this is a south (S) bearing, the product (16.148) of multiplying distance times cosine is listed under south.

The measured distance, 165 feet, is multiplied by the sine of the measured angle 84°23'. This gives you 164.207 feet, the east departure. Note again that the bearing is south by east.

To balance the survey, total the columns under latitudes and departures. If the north and south columns are equal, then the latitudes balance. The same is true for departures under the east and west columns. These columns balance in Figure 3-2.

The DMD. Listed under D.M. Distance (DMD) is the *double meridian distance* figures. This is perhaps the most difficult step of the computations. It is simpler divided into three steps:

1. The DMD of the first course (course *AB* starting at point A on the north-south reference meridian) equals the departure of the course itself (164.207 feet).

2. The DMD of any course equals the DMD of the preceding course, plus the departure of the preceding course, plus the departure of the course itself.

3. The DMD of the last course should numerically equal its departure, but with the opposite sign. Note that east departure is positive and west departure is negative. Thus, since survey calculations begin at the westernmost point of the survey, the departure of the first course is easterly and positive.

Starting with step 1 above, *AB* is the first course and the distance 164.207 feet is both the departure and the DMD.

Then, step 2 computes the DMD for course *BC*. The DMD of any course (*BC*) equals the DMD of the preceding course (164.207 feet, course *AB*), plus the departure of the preceding course (164.207 feet, course *AB*; simply multiply by 2), plus the departure of the course itself (25.669 feet, course *BC*). So, 2 × 164.207 feet + 25.669 feet 354.083 feet, the DMD of course *BC*.

Using step 3, the DMD of course *CD* equals the DMD of the preceding course (354.083 feet, course *BC*) plus the departure of the preceding course (25.669 feet, course *BC*) plus the departure of the course itself (negative 164.207 feet, course *CD*). Course *CD* has a west bearing which is negative: 354.083 + 25.669 − 164.207 feet = 215.545 feet, the DMD of course *CD*.

And again, using step 2, the DMD of course *DA* is the DMD of course *CD* (215.545 feet) plus the departure of course *CD* (negative 164.207 feet) plus the departure of course *DA* (negative 25.669 feet): 215.545 − 164.207 − 25.669 feet = 25.669 feet, the DMD of course *DA*.

The Double Areas. In the columns of Double Areas, north is positive and south is negative. You find the double areas by multiplying the DMD of each course by the latitude of the same course. For course *AB* the double area is 164.207 feet (DMD) × (-16.148 feet negative latitude) = − 2651.614 square feet.

For course *BC* the double area is 354.083 feet (DMD) × +275.466 feet (positive latitude) = +97,537.827 square feet.

Find the double areas for courses *CD* and *DA* the same way. Then total and add the double area columns: (+101,018.447) + (− 9,722.550) = 91,295.897 square feet. Because this is a double area you divide by 2, which gives you 45,647.948 square feet, the actual area in square feet of the survey.

■ Finding the Area

The symbol (÷k) is the constant number of square feet in an acre. If you divide the number of square feet in the survey (45,647.948) by the number of square feet in an acre (43,560), you get 1.0479 acres (or 1.048), which is the amount to be listed in the deed.

Economizing in the Field

■ A Sample Survey

You should organize your fieldwork so that you need to make as few transit setups and distance measurements as possible. The following is a well-organized survey.

In Figure 3-4 the 1/4-1/4 line runs due north. It is simple to set up on the 1/4-1/4 line and sight in on it. Then turn a 90° angle to a due west bearing.

By doing this, the line of sight directed west may or may not strike the fence-corner post at point *C*. (The fence line is supposed to be the north line of the survey.) You need to find a point from which the fence post lies due west. To do this, establish the 1/4-1/4 line to a point south of point C.

In Figure 3-4 this is point Y. If the area is rough or covered with trees, check an aerial map at the county engineer's office to find some distinguishing feature along the 1/4-1/4 line south of point C to survey to and mark it point Y.

Set up on point Y and direct the line of transit sight north along the 1/4-1/4 line and turn a left angle to the fence post at point C. Assume this angle is 84°17'21.86".

Measure the distance from Y to C (301.496 feet). Then, find the length of CX.

$$\text{Sin } Y = \frac{\text{side opposite}}{\text{hypotenuse}} = \frac{CX}{301.496}$$

CX = sin Y times 301.496 = .99483 × 301.496 = 300 feet.

Find the length XY and establish point X.

$$\text{Cos } Y = \frac{\text{side adjacent}}{\text{hypotenuse}} = \frac{XY}{301.496}$$

XY= cos Y times 301.496 = .0996037 × 301.496 = 30 feet.

Point C is due west of point X. Measure from point X to the reference point (the 1/4-1/4 corner), and you get 500 feet, the measurement for the deed description.

To double-check, set up on point X and turn 90° off the 1/4-1/4 line. The line of sight due west should center on the fence post. While you are at point X set a point in the due west line about 15 feet east of the fence post at C. Use this point to find the bearing of the fence line running northwest (point m, Figure 3-4).

Next, set up on point m. Take a backsight to point X and flop the telescope for a due west sight. Set point n on a due west line 100 feet from the center of the fence post at point C (the 100-foot distance is arbitrary).

Then, set up on point n, line in on point m, and turn a left 90° angle. Set point o in the fence line between the centers of adjacent fence posts. Measure line on. Compute the angle for oCn. Line on measures 9.83 feet. Thus,

$$\text{tan angle } nCo = \frac{on}{nC} = \frac{9.83}{100} = 0.0983$$

arc tan 0.0983 = 5°36'51" (the value of nCo)

The bearing of the fence line is 90° − (5°36'51") = 84°23'9". And because you carried this bearing from the true north bearing of the 1/4-1/4 line, it is a true bearing.

At the Drawing Board. As you learned, in Figure 3-1 the lettering (point A) of the drawing is such that the first course for survey computations begins at the Westernmost corner. The order of progress from point *A* makes the east departures positive and the west departures negative. In an actual survey shown in Figure 3-5, the westernmost corner (point *A* at the southwest corner of the survey) is the beginning point for your office computations. The courses used for computations travel consecutively around the survey in a counterclockwise direction. Point *A* is located at the intersection of two road intersections.

The beginning point for the fieldwork, as given in the legal description, is point *E*. This point is the northwest corner of the SE 1/4 of the SE 1/4. This beginning point is a legal, political point from which to begin the survey and the description.

The points *A, B, C, D,* and *E* appear only on a drawing for office use. They are your reference but do not go on the survey map.

The courses, bearings, and distances are listed in Figure 3-5. The cosine and sine of each bearing are listed only for instructional purposes in Figure 3-6. Normally you see them only on the calculator; they aren't recorded this way.

The bearing of course *AB* is S79'38'E. The cosine and sine of this bearing are, respectively, .17995 and .98367. Because this is a south bearing, the distance multiplied by the cosine gives you a south latitude of 26.9900. Put this figure in the south column. The bearing also bears east and the distance times the sine yields a departure of 147.5500 feet. Put this figure in the east column.

Compute the latitudes and departures and list them in the proper columns for each course. Then total the columns and subtract the smaller total from the larger. This is shown in the north column of the latitudes and in the west column of the departures. There is a small error in the survey balance that is shown in each column.

Add the course distances to get a total of 1167.21 feet. You learn how to balance the survey in the following sections.

■ Errors of Closure

Figure 3-7 shows that this survey has an error of closure of 0.1768 foot. The computations began at point *A* and ended at point *A'*. The course *EA* failed to return to point *A* by 0.1768 foot. It ended 0.122 foot north and 0.128 foot west of the starting point *A*.

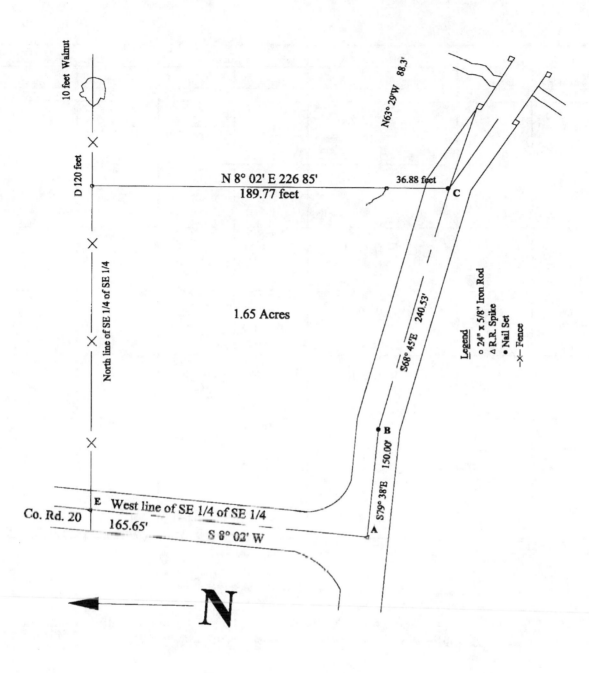

Figure 3-5 *Survey computations begin at the westernmost corner*

Course	Bearing	Distance		Latitudes		Departures	
		Cosine	Sine	North	South	East	West
AB	S 79°38' E	150 .17995 150	00 .98364 02		26.9900 + 0.0056 26.9959	147.5500 + 0.0234 147.5734	
BC	S 68°45' E	240 .36244	53 .93201		87.1770 + 0.0191 87.1961	224.1760 + 0.0355 224.2115	
CD	N 08°02' E	226 .99018	85 .13975	224.6230 - 0.0492 224.5742		31.7020 + 0.0050 31.7070	
DE	N 81°58' W	384 .13975	18 .99018	53.6890 - 0.0117 53.6773			380.4070 - 0.0603 380.3467
EA	S 08°02' E	165 .99018	65 .13975		164.0230 + 0.0359 164.0589		23.1490 - 0.0036 23.1454
		1167	21	278.312 278.190 ———— 0.122	278.190 278.312 ———— 556.502	403.428 403.556 ———— 806.984	403.556 403.428 ———— 0.128

Figure 3-6 *Calculations for survey*

The sum of the squares on the two sides of a triangle is equal to the square on the hypotenuse of the triangle. So,

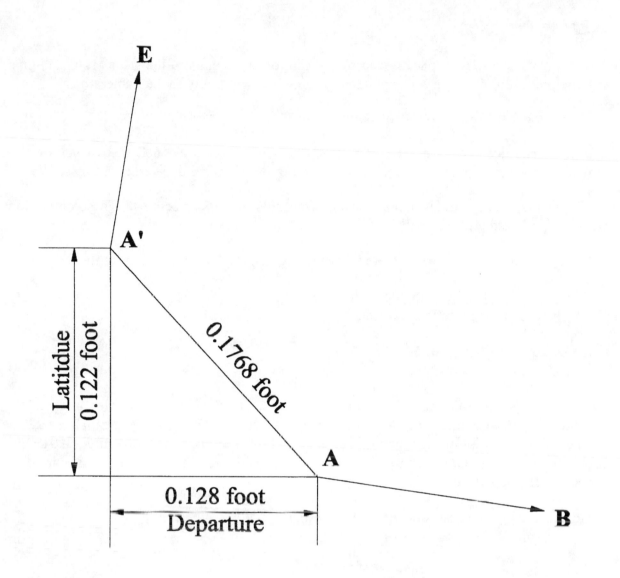

Figure 3-7 *Calculating the error of closure*

Error of closure = $\dfrac{\sqrt{(0.122)^2 + (0.128)^2 =}}{\sqrt{(0.014884) + (0.016384)}}$ = 0.1768

Precision = $\dfrac{0.1758 \text{ (Error of closure)}}{1167.21 \text{(Sum of all sides)}}$ = $\dfrac{1}{6602}$

The square root of the square on the hypotenuse of a closure triangle is the error of closure. So,

$\sqrt{0.031268}$ = 0.1768, the error of closure

■ Survey Precision

The error of closure (0.1768) divided by the sum of the lengths of the sides (1167.21) gives you the precision of the survey expressed by a fraction with unity as the numerator.

Precision = 0.1768/1167.21 = 1/6602

This is an acceptable survey. There is 1 foot of mistake in 6602 feet of survey line. However, for instructional purposes, the procedure for balancing this survey follows.

To balance the latitudes and departures, you must make corrections. These corrections will be distributed proportionally among all sides of the survey.

The corrections for course *AB* are shown in Figure 3-6. The correction for the latitude is:

$$\frac{\text{total error in latitude} \times \text{latitude of the side}}{\text{sum of the latitude of all sides}}$$

$$= \frac{0.122 \times 26.99}{556.502} = 0.0059.$$

Since the south column has a smaller sum (278.190) than the north column (278.312), add the correction to the latitude of course *AB* to get 26.9959. (The correction would be subtracted in a column with the larger total. For example, the correction, 0.0492, is subtracted for course *CD* since the north column has the larger total, 278.312.) The correction for course *AB* departure is

$$\frac{\text{total error in departure} \times \text{departure of the side}}{\text{sum of the departures of all sides}}$$

$$= \frac{0.128 \times 147.55}{806.984} = 0.0234.$$

The correction (0.0234) is added because the east column has the smaller total.

Notice that the changes in the latitudes and departures have created changes in the lengths and bearings of the sides.

$$\text{corrected length} = \sqrt{(\text{latitude})^2 + (\text{departure})^2}$$
$$= \sqrt{(26.9959)^2 + (147.5734)^2} = 150.02$$

Mark the new distance in the Distance column, as shown. The corrected bearing is

tan bearing = departure/latitude = 147.5734/26.9959
= 5.46651. Arc tan = 79°38'.

All corrections to latitude and departure are given in Figure 3-6. As mentioned, you usually do not need the corrections to latitudes, departures, distances, and bearings for a survey that initially balances as well as this one did. But, you do need to know how to do these calculations because you will need to make corrections when the error of closure produces a precision lower than the standard set for a given area (see also, Chapter 12). Find out what the local standard is from the county engineer.

All essential information about the survey is given in Figure 3-5. But since there are various requirements for each local area find out what's needed from the county recorder or county engineer.

CHAPTER 4

Using a Transit and Tape

The transit and tape are the basic instruments you will be using in the field. This chapter gives you methods for using the transit and tape to measure a variety of complex lines and curves.

Reference Lines

As mentioned previously, you should always reference at least one line of a survey. You can do this in various ways, but the best is to use a true north line that has already been run close to the site, or to establish a new true north line at this location (see Chapter 15).

You should monument two points in the true north line. To do this, set up the transit over one point, turn an angle from the other point, and reference a point in one line of the survey. Repeat this with an angle turned to another point in the survey line.

Then, measure the distance from the transit point to each point in the survey line. You should also monument the survey-line points using 5/8-inch re-bar or concrete. Remember that a survey line must be referenced to a permanent point by an angle and measurement. The best point for reference is a National Survey point.

If you carefully monument each end of a survey line with two permanent points, you can reference the line by its bearing from true north. Some surveyors reference only one point in a survey. One point is not enough. It is only a pivot point from which to start a survey in any direction. Be sure to use two points. Define and mark them carefully, and make them relatively permanent.

■ Finding the Intersection of Lines

To find the intersection of two straight lines, set up the transit over a point on one line and within a few feet of the point of intersection of the two lines. Set two stakes in this line, one on each side of the point of intersection. Then place a mark on each stake exactly on the line as shown in Figure 4-1.

Next, set up over a point in the second line and sight in the line. Stretch a cord between the marks on the stakes previously set. Sight the second line in on the chord. Drop a plumb bob (a metal bob) from the point of intersection on the line. Then remove the cord and drive a stake at the point of intersection.

Now, replace the cord between the stakes and again sight in the second line on the cord. Transfer this point by plumb bob to the stake below. You have found the point of intersection. Mark it with a surveyor's tack.

■ Defining Curves in the Field

When you are in the field you will run into situations requiring you to lay out circular courts, turn-arounds, and curves between intersecting streets. In such cases you need to know how to place an arc of a circle between two straight lines that are tangent to the arc. The following situations show you how to approach various problems (see also Figures 10-9 and 10-10).

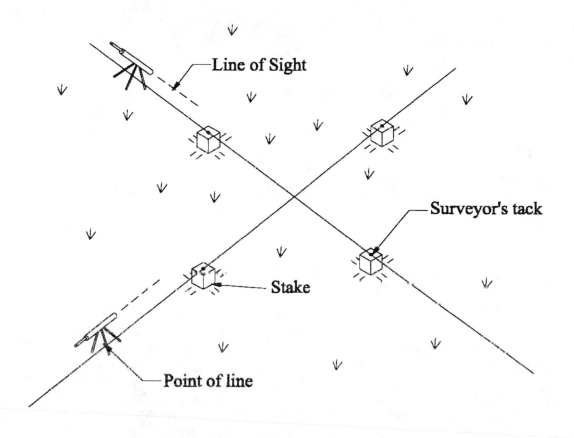

Figure 4-1 *Finding the intersection of two lines*

Assume you have a street situation that resembles that in Figure 4-2. The arc (curve) is placed such that the straight-line streets are tangent to it. In ordinary street construction the radius of a curve is comparatively small and you can find the points on the arc by swinging a tape. Suppose that in this case, the radius is too large; you cannot use a tape. Instead, use math and a transit.

As in other curves, start working from the point of curve (*PC*) to the point of tangent (*PT*). In Figure 4-2 the tangents *AG* and *BG* intersect at point *G* by prolonging the two lines between which the arc (*PC* to *PT*) is drawn.

Then, *PC* to *G* = *PT* to *G* (tangents labeled *T*). At point *G* angle *I* is the central angle (1/2*I* = 1/2*I*), and *PC* to 0 *PT* to 0 = R, the radius of the curve. *PC* to 1, 1 to 2, and 2 to 3 are the chords used in the fieldwork (labeled *C*), which are about 100 feet long (chords must be equal), or whatever length is suitable.

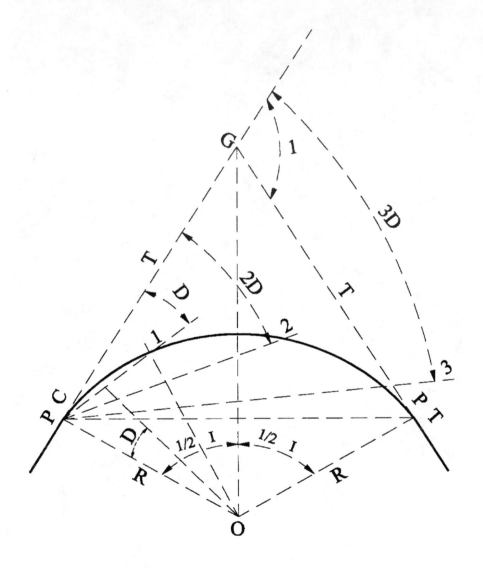

Figure 4-2 *Placing the arc of a circle*

G-PC-1 is the deflection angle *D*. The angle turned on point *PC* from *G* to 2 = 2*D*, and from *G* to 3 = 3*D*.

Do the following calculations based on geometry and trigonometry.

G-PC-O = G-PT-O = 90°

G-PC = G-PT, and G-PC-PT = G-PT-PC = 1/2I

I = PC-O-PT, and PC-O-G = PT-O-G = 1/2I

G-PC-1 = 1/2 (PC-O-1) = D

G-PC-2 = 1/2 (PC-O-2) = 2D

G-PC-3 = 1/2 (PC-O-3) = 3D

Because the chords are equal, PC-O-1 = 1-O-2 = 2-O-3. Chord 3-PT is not a full chord and therefore is a subchord. Then 3-PC-PT is a subdeflection angle.

Follow these formulas for calculations.

Given I and T, find R: $R = T \times \cot 1/2I$.

Given R and I, find T: $T = R \times \tan 1/2I$.

Given R and C, find D: $\operatorname{Sin} D = 1/2C/R$. $D = \arcsin$ of sin D.

Given C and D, find R: $R = 1/2C/\sin D$.

Given I, D, and C, find T: $T = (C \times \tan 1/2I)/2\sin D$.

Given I, T, and C, find D: $\operatorname{Sin} D = (C \times \tan 1/2I)/2T$.

If 100-foot chords are used:

$\operatorname{Sin} D = 50/R$; $R = 50/\sin D$.

$\operatorname{Sin} D = (50 \times \tan 1/2I)/T$; $T = (50 \times \tan 1/2I)/\sin D$.

Remember that the number of chords in the length of a curve is N = 1/2I/D. If D is not evenly divisible into 1/2I, there will be a subchord, as mentioned above. But generally, the length of curve is the total of the chord lengths instead of the length on the curve.

In Figure 4-2 there are three full chords and a subchord. Each chord is the standard 100-foot length which with the subchord, is 3.624 separate lengths as found by N = 1/2I/D. The length of curve is 362.4 feet.

You can establish the degree of curve by the number of degrees at the center (O) subtended by a 100-foot chord. That is, if PC-1 is a 100-foot chord, and PC-O-1 is an angle of 5°, the degree of curve is 5° and the curve is called a 5° curve.

The curve is a 4° curve when *PC-O-1* is 4°, and 2° when *PC-O-1* 2°, because the chords are all 100 feet long.

Now that you know the number of chords and the length of the curve, consider the problem: How do you connect the two streets by the arc of a circle tangent to one street at a point *PC* and tangent to the other street at some point *PT* to be established? First, in the field, start by prolonging the centerline of the streets until they intersect at some point *G*. Measure angle *I* with the transit.

Second, measure *PC-G (T)* and measure from points along the other street centerline to place point *PT* so that *PC-G* equals *G-PT* (both labeled *T*). These are the tangents.

Third, use the formula sin $D = (C \times \tan 1/2I)2T$ to calculate the deflection angle *D*. Turn the deflection angles and stake them accurately. Set up on *PC*, backsight on *G*, and turn angle *D*. Hold the tape end at *PC* and set point 1 in the line of sight 100 feet from *PC*.

Again, backsight on *G* and turn angle 2*D*. Set point 2 in the line of sight 100 feet from point 1. And, in the same way turn angle 3*D* and set point 3.

Fourth, using $N = 1/2I/D$, find the length of curve and subtract 300 to get 62.4 feet (from above, 362.4 − 300 = 62.4 feet). This is the length of the subchord.

Finally, measure 62.4 feet from point 3 to determine the *PT*. This point should correspond with the *PT* set by measurement from point *G*.

Possible Problems. If the situation is such that you cannot see all points on the curve from a setup on *PC*, move the transit and set up on the last point on the curve set from point *PC*.

Assume point 2 is the last point that you can set from *PC*. Set up on point 2, backsight on *PC* with the scales set at zero (the deflection at *PC* is zero), plunge the telescope, and turn an angle equal to 3*D*. The deflection for point 3 is 3*D*. Now, set point 3, 100 feet from point 2.

If the point *PT* is not visible from point 2, set up on point 3, back sight on point 2 with the scales set at 2*D* (the deflection for point 2 is 2*D*), plunge the telescope, and turn an angle equal to 1/2*I*, because the deflection for *PT* is 1/2*I*.

Whenever you are sighting at any point on a curve, the scales must read the total deflection angle used for locating that particular point. When you are sighting along a tangent to the curve at any set point, the scales should read an angle equal to the total deflection angle for that point.

■ Situation 2

You can use a tape to lay out small circular curves by measuring out each deflection angle with a measurement from the tangent (an extension of the street to which the curve is being fitted). This method is shown in Figure 4-3 in which the angle $EA1$ is the deflection angle. $E1$ and $F2$ are measurements from tangent AY. AI is a 100-foot chord, and $CO1 = D$. You need to find offset $E1$. In this situation it is best to work with decimal degrees instead of converting to degrees, minutes, and seconds.

Start by recognizing that from geometry a radius is perpendicular to a chord, and then angle $OC1 = 90°$. Also, from geometry, triangle $OC1$ is a right triangle. From trigonometry, sin angle $CO1 =$ side opposite/hypotenuse $= C1/O1$. And, line $C1 = 1/2$ chord $= 50$ feet. Thus,

Sin $CO1 = 50/400 = 0.125$.

Angle $CO1$ equals arcsin $0.125 = 7.18°$.

Angle $CO1 = D$, then $D = 7.18°$.

Angle $BO1 = 2D$. $2 \times 7.18° = 14.36°$.

Sin $14.36° = B1/1400$;

and $B1 = 0.24801 \times 400 = 99.205$ feet.

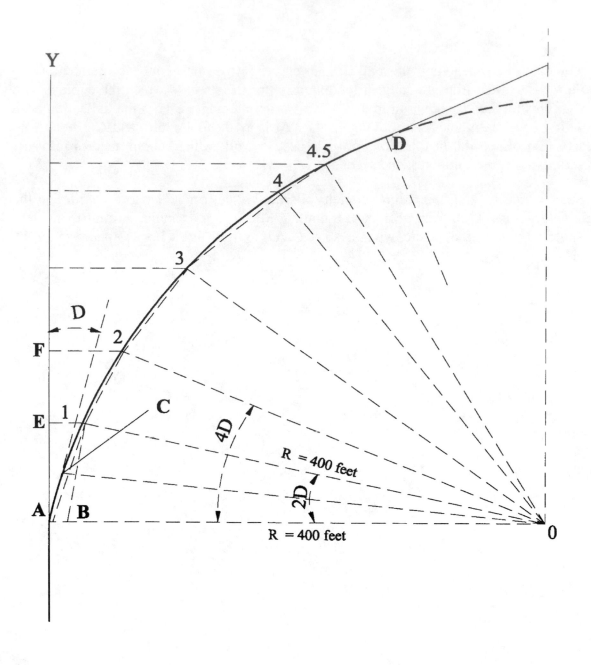

Figure 4-3 *Measuring deflection angles*

In rectangle *AE*1*B*, chord *A*l bisects angles *E*1*B* and *EAB*. Side *AE* = side *B*1 and the two sides are parallel. Also, sides *E*l and *AB* are equal and parallel.

Chord *A*1 = 100 feet. Side *B*1 = 99.205 feet. Because angle *EA*1 = *D* = 7.18°, angle *A*1*B* = *D* = 7.18°. [By geometry, when a line, *A*1, bisects two parallel lines, *E*l and *AB*, the alternate interior angles are equal (angle *EA*1 = angle *A*1*B*)].

Sin angle $A1B$ = side opposite/hypotenuse,

$\sin A1B = AB/A1$,

$0.125 = AB/100$, and

$AB = 0.125 \times 100 = 12.50$ feet.

Thus, because AB equals $E1$, the offset $E1$ equals 12.50 feet.

Similarly, you can solve the triangle formed by a perpendicular dropped from point 2 to radius AO. Then you can solve the rectangle formed by dropping the perpendicular, which will be parallel to line AY, to find the offset from point F to 2. For this problem, from point 2 the angle will be $4D$, or $4 \times 7.18° = 28.72°$.

Using a tape for layout to find point 1, lay out AE. Pin a tape at A and a tape at E. Stretch both tapes toward point 1 until 12.5 feet ($E1$) and 100 feet (chord $A1$) meet on the tapes. This is point 1.

Second, after you have computed the length of the offset for point 2, measure that length from point F to match a 100-foot length (chord 1-2) from point 1 (to set point 2). Finally, you will find the point of tangency, PT, at the end of a 50-foot chord. The distance is point 4 to point 4.5.

Laying Out Roads

You will often need to lay out simple roads that do not require extreme accuracy for parks, long driveways, and landscaping areas. To do this start by getting a map of the area showing buildings and other features to be serviced by the road. Scale the distances from an assumed baseline on the map to each feature on the map to be serviced by the road.

Next, plot this information on a larger-scale drawing with an accuracy consistent with the measurements required for layout in the field. Draw the baseline first and plot from there.

Laying Out a Line

Occasionally, you will need to set up when the terrain is such that a transit setup must be made on a line between two points, neither visible from the other. To do this, first have two assistants establish the line as well as possible by aligning range poles.

Figure 4-4 shows this situation. The points are shown as X and Y with range poles set on the points. An assistant with a range pole stands at point A, which is as close to point X as possible while still within sight of the range pole at point *Y*. Another assistant with a range pole stands at point *B*, which is as close to point *Y* as possible and while still within sight of the range pole at point *X*. In Figure 4-4 these four points are projected as the black dots with a straight line between dots *X* and *Y*. The broken lines represent the line of sight of each assistant.

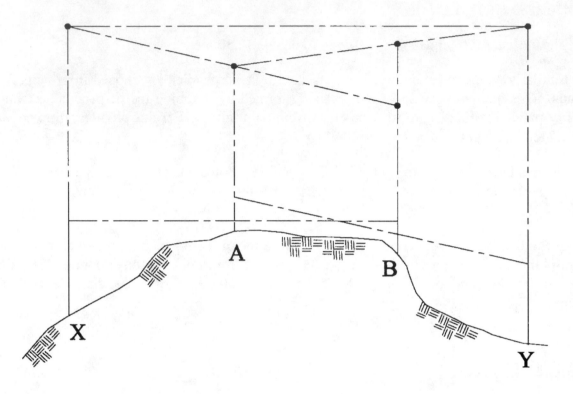

Figure 4-4 *Aligning range poles*

Assistant B lines in assistant A with point *A*. Then assistant A lines in assistant B with point *B*. This action is continued, moving each assistant, action by action, until they are closely aligned with points *A* and *B*.

Set up a transit on this line at a point between *A* and *B* and take sights alternately between *X* and *Y*. Move the transit, if necessary, to place it exactly on a true line between points.

Often, you will need to measure the angle at a fence corner. Heavy growth of brush, briars, or timber can make this a difficult task. Figure 4-5A shows a sample situation.

The corner is at *B*, and *AB* and *BD* are portions of the fence lines tied to corner *B*. Set points *A* and *D*, each at 20 feet from *B*. Measure the distance *AD* as 37.2 feet. One half of *AD* is 37.2/2 = 18.6 feet. Then,

sin angle *ABC* = side opposite/hypotenuse

$= AC/AB = 18.6/20 = 0.93$.

Arcsin $= 68°26'05"$, angle *ABC*. And since

angle *DBC* = angle *ABC*,

angle *ABD* $= 2 \times 68°26'05" = 136°52'10"$.

The deflection angle from line *BE* to the fence line is $180° - (136°52'10") = (179°59'60") - (136°52'10") = 43°07'50"$. This is an easy way to subtract or add degrees, minutes, and seconds.

In Figure 4-5B the angle at fence corner *B* is needed. Assume trees and shrubs prevent you from directly turning the angle with an instrument. There is no way for you to measure the east line (*BC*), but you need both a distance and bearing for line *BC* in order to compute angle *ABC*.

You can run a traverse line around the growth from corner *A* and back to corner *B*. Measure the distance *AB* at 400 feet. Set point *E* in the west-line fence 400 feet from point *A*. Check the line *AE* for bearing and you will find it runs N20°30'E from true north.

From *E* find a clear sighting to point *D* (any clear space) and turn an angle R (right) 95°20' off *AE* to point *D*. The line *AE* lies in quadrant I of the bearing chart (see Figure 1-4). If you turn R95°20' off a bearing at N20°30'E, the new bearing will be in quadrant IV.

Because you turned the bearing N20°30'E to the right off true north, and turned angle 95°20' right off bearing N20°30'E, the total angle to the right from true north is $(95°20') + (20°30') = 115°50'$.

Because quadrant I is 90°, the amount of angle into quadrant IV is $(115°50') - 90° = 25°50'$. And because you referenced the bearings in quadrant IV from a true south direction, $90° - (25°50') = (89°60') - (25°50') = 64°10'$. Thus the bearing for *ED* is S64°10'E.

The corner angle at *A* is 90°. Then, $90° + 20°30' = 110°30'$. The bearing of line *AB* lies in quadrant IV. If you subtract, $110°30' - 90° = 20°30'$, an angle in quadrant IV. Then, from the 90° in quadrant IV subtract 20°30' to get a bearing on *AB* of $90° - (20°30') = 69°30'$, or S69°30'E (see Figure 4-5C).

You can now compute the bearing and distance of line *BC*. Figure 4-6 shows the computations. You may recognize the chart from the examples in Chapter 3. And although that chapter discusses methods of survey closure, the following section is included for further practice (see also Chapter 12).

In Figure 4-5B, all distances and bearing are given except those for *BC*. List this information in Figure 4-6 under Course and Distance. Also, find and list the cosine and sine of each bearing as shown.

The direction of listing the courses is alphabetical, *AB*, *BC*, *CD*, etc., beginning with the westernmost point of the survey. When courses *AE*, *ED*, and *DC* were sighted and measured, the work was done clockwise. In Figure 4-6, the measurements are listed counterclockwise.

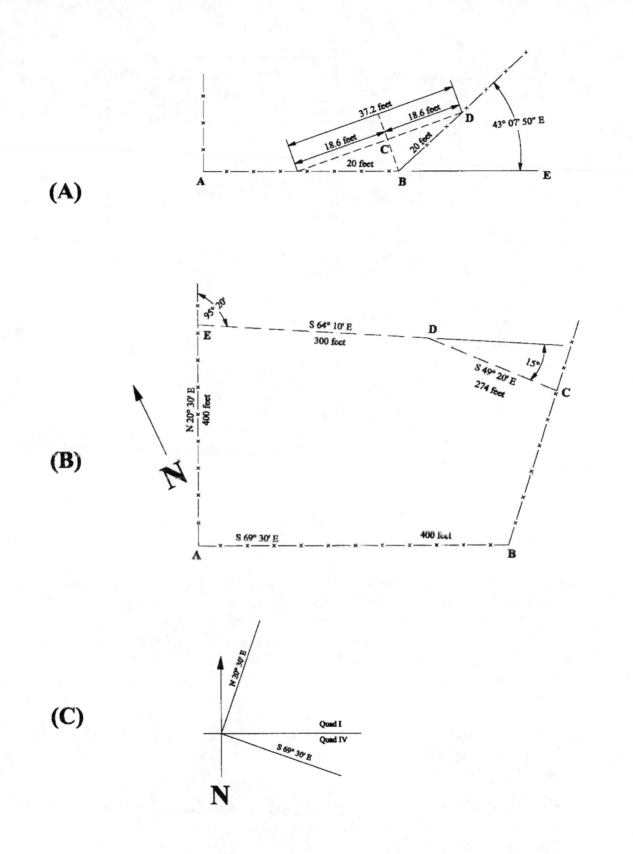

(A)

(B)

(C)

Figure 4-5 *Charting the survey*

Course		Distance		Latitudes		Departures	
No.	Bearing	Cosine	Sine	North	South	East	West
AB	S 69°30' E	400 .35021	00 .93667		140.082	374.669	
BC	N 49°48'51" E	318 Tan = 1.18393	427	205.471		243.265	
CD	N 49°20' W	274 .65166	00 .75851	178.5547			207.832
DE	N 64°10' W	300 .43575	00 .90007	130.726			270.019
EA	S 20°30' W	400 .93667	00 .35021		374.669		140.082
				309.280 + 205.471	514.751 - 309.280	374.669 + 243.265	617.934 - 374.669
				514.751	205.471	617.935	243.265

Figure 4-6 Closing the survey

For example, course *DC* on the drawing is S49°20'E. The back bearing of this course is *CD*, N49°20'W, as listed under Course. For this bearing, you must list the product of the distance and the trigonometric function under North and West. Try course *AB*. The cosine of 69°30' is .35021, which multiplied by the distance gives you .35021 × 400 = 140.082. List this under South, the primary direction of the course. The sine of 69°30' is .93667, which, multiplied by the distance, gives you .93667 × 400 = 374.669, the secondary direction of the course.

Courses *CD*, *DE*, and *EA* are listed in the appropriate columns under latitudes and departures. Since nothing is listed for course *BC*, the column totals are (from left to right), 309.280, 514.751, 374.669, and 617.934.

As explained in Chapter 3, the latitudes and departures must balance to bring the survey traverse back to point A. The difference in the north and south latitudes is 514.751 − 309.280 205.471. The north column for total latitude is 309.280 + 205.471 = 514.751. Distance 205.471 is the latitude for course *BC*.

Under Departures, column West is larger, so 617.934 − 374.669 − 243.264. Place this under East for the *BC* departure and add the column for the total departure east, or 617.934.

The length of the course is:

$$BC = \sqrt{\text{latitude}^2 + \text{departure}^2} = \sqrt{(205.471)^2 + (243.264)^2} = 318.427.$$

Write this in under Distance.

The tangent of the bearing equals departure/latitude, 243.264/205.471 = 1.18393, arc tan 1.18393 = 49.81415812°, or 49°48'51". Counterclockwise, line *BC* travels north by east, and the complete bearing is N49°48'51"E. To convert 49.81415812° to degrees, minutes, and seconds, the digits before the decimal are subtracted (49°). Multiply the remainder by 60 to get (0.81415812×60) 48.84948720. The digits here before the decimal are in minutes. The reading now is 49°48'. Again, multiply the decimal part: $0.84948720 \times 60 = 50.96923200$. This rounds off to 51", so the full reading is 49°48'51". The bearing is N49°48'51"E.

These computations were done with a scientific calculator, as you can tell by the long decimals. This method gives you the most accuracy. You can round off to practical units after the final calculation. For instance, you could state the bearing as N49°48'E without appreciable error.

Remember to run as few lines as possible to get results. If you could have sighted a line and measured from point *A* to point *C* with a bearing and distance for both *AB* and *AC*, then you could have computed line *BC* in the same way.

Setting Parallel Lines

■ Using Transit and Tape

A practical method to establish a line parallel to a given line using both transit and tape starts when you set up the transit 6 to 10 feet from the line. To find the shortest distance from the line to the transit, extend the tape from the line to a point in front of the transit. Then swing the tape in a horizontal arc. The shortest distance read on the tape by your instrument assistant is the actual distance from the transit's center to the line.

Set one point by plumb bob under the transit and another at any convenient distance from the transit. The line between these points is parallel to the given line.

■ Using Tape Only

To establish parallel lines with a tape only, measure as shown in Figure 4-7. Using multiples of the 3-4-5 triangle, set distances at 9-12-15 to establish a point (*F*) on the parallel line.

Pin points *B* and *C* 24 feet apart on the existing line *XX*. Then pin a tape end at *B* and match 15 feet on it with 9 feet on a tape from point *F*. Do the same from point *C*. By the principles of geometry, triangle *ABF* equals triangle *ACF*, and angles *AFB* and *AFC* are each 90°. If you measure *BD* and *CE* at 9 feet and *AD* and *AE* at 12 feet, points *D* and *E* are set in line with point *A*. A line connecting these three points is parallel to line *XX*.

Figure 4-8 shows a 400 × 450 foot rectangular lot with a proposed building as shown. Staking the lot is a simple job. Check out the land area to see if it conforms with the land deed. If this is a parcel to be parted off from farmland on the town's outskirts, there still isn't a deed at all. Use the farm deed and corners of the farm to get started.

If the lot lies between other properties, the corner markers might already be in place. Check the lot lines to see if the dimensions are correct. Sometimes surveys on adjacent lots are incorrect and markers might be placed that reduce the area in your lot.

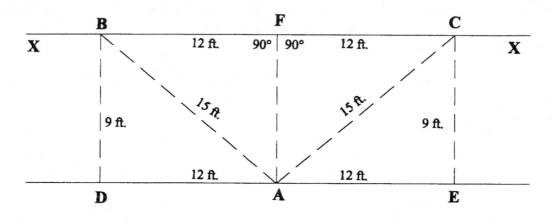

Figure 4-7 Establishing parallel lines

Note that the building is composed of two sections, a 100 × 360 foot warehouse and a 200 × 270 foot sales room. Batter boards are installed at each corner in spots convenient to construction. Lines are stretched from board to board intersecting to form courses of the exact dimensions of the building. Note also that certain dimensions from the lot lines are maintained and building lines run parallel to lot lines. Therefore, set points corresponding to the proposed layout along the lot lines. Begin taking measurements at the rear lot corner *R*, the primary reference point.

First, set up on R and set stakes 1, 3, 5, and 7. Place a surveyor's tack in each stake exactly on the lot line. Turn the transit sight on the front corner, and set points 9, 11, and 13 (with both stakes and tacks).

Second, set up on R', sight on the front corner, and set points 10, 12, and 14. And, set up on R'' and set points 2, 4, 6, and 8. Set stakes and tacks at every point around the lot.

Third, set up on point 1, sight on point 2, and then mark building line points on the batter boards at C and D. From points 7 and 8 set marks A and B. Repeat this from point to point until all the boards are marked. Along the side lines set marks from 9 to 10, 11 to 12, and 13 to 14.

Diagonals, shown in the figure from mark to mark as broken lines, should measure as equal for the building lines to be square. Unless you take these measurements electronically, be sure to use intermediate supports for the tapes. If the marks do present a square pattern, tautly stretched small-gauge wires will lie exactly on the building lines.

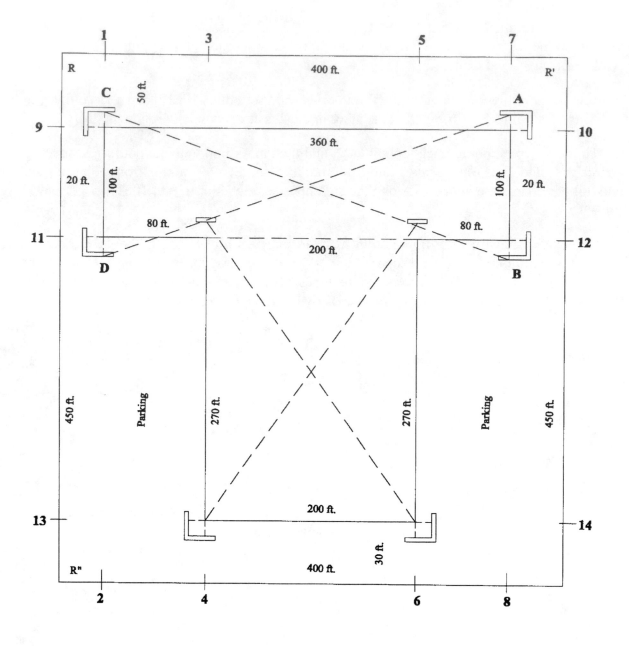

Figure 48 *A 400 × 450 foot rectangular lot*

A plumb bob dropped from each intersection of the wires, such as point *X*, transfers the corner point from the wires to the ground for foundation work, form building, or to mark the corner point to begin masonry work. These are the *masonry lines*.

Planning A Bridge

Assume that a bridge is required on the property you are surveying. You will need to plan lines for a wing abutment. If piers also are required, you can show the work along with that for the abutment.

As shown in Figure 4-9, first establish the centerline of the bridge (called the *axis* of the bridge). Generally, this is an extension of the road centerline.

Second, set at least two points in the road centerline 200 feet apart. Put the first point 50 feet from the proposed stream face of the abutment. These are reference points only. They are intended as baselines for all other lines and points.

Third, set Station 12 + 50 on the road centerline. This Station marks the stream-side wall face of the abutment. Then set up on station 12 + 50 and set point 1 and 2 at 90° to the road centerline. From this setup establish points 3 and 4 where the wing walls join the front wall.

Now you can set up on point 3, and then point 4, to turn off the angles of the wing walls. Set points 5 and 6. These are all the points you need for form building.

> **Note:** Streams generally have stream banks on which you can put Station 12 + 50 and points 1, 2, 3, and 4. However, if the wall face falls in the stream, then you will need to set offset stakes between Station 12 + 00 and Station 12 + 50. Unless, of course, there are piles or a cofferdam to place them on. If offsets are necessary, try to make the distances some whole number of feet.

After you have staked out the preliminary layout and completed the plan of action for form building, set the batter boards. Set all points at whatever distance is convenient to construction under the prevailing stream and bank conditions. Transfer all markings and elevations to the batter boards.

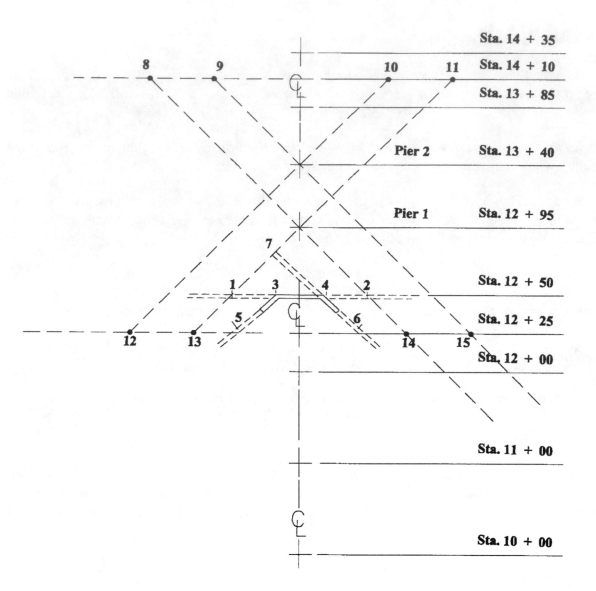

Figure 4-9 *Setting points for a bridge*

You might need a point such as point 7. If there's nothing to place it on, and the stream is too deep to place it there, you can place it on the opposite bank.

The pier centers will lie on the bridge centerline. The piers stand at Stations 12 + 95 and 13 + 40. Station 13 + 85 is on the streamside face of the opposite wing abutment. Establish a baseline on each stream bank, one at Station 12 + 25 and one at Station 14 + 10. Set points 8, 9, 10, 11, 12, 13, 14, and 15 on the baselines, with the points at the ends of lines intersecting on the pier centers. Choose some standard of angle that gives you an easy computation, for instance, 45°. Because pier 1 stands at Stations 12 + 95, 70 feet from the baseline at Station 12 + 25, point 14 lies on the baseline 70 feet from the centerline. This establishes the 45° angle because the short sides of a 45° triangle are equal.

Points 9, 10, and 13 also lie 70 feet from the baseline. In the same way, points 8, 11, 12, and 15 are set 115 feet from the baseline. From Station 12 + 25 to Station 13 + 40 is 115 feet, and the distance from Station 12 + 95 to Station 14 + 10 is also 115 feet.

After these points are set, you only need to sight from one to the other along the 45° diagonals, as shown, to find the pier centers. You do not need to turn any other angles.

CHAPTER 5

Applied Geometry Using a Transit

Every shape is either a geometric figure or a combination of geometric figures. In surveying you need to know how to construct and measure geometric figures both in the field and on maps. At the drawing board you use protractors and pencils; in the field you use compasses, transits, tape measures, and non-stretch cords.

This chapter explains how to lay out various geometric figures and gives examples of situations you might encounter in the field. Generally, you will want to work out the problem on paper first, then approach it in the field.

For instance, suppose you are at a site faced with a survey that looks like Figure 5-1. The line YY bisects line XX at point O. Points A and B are at equal distances from the ends of line XX. You measure and find out that the distances AC, AD, BC, and BD are all equal.

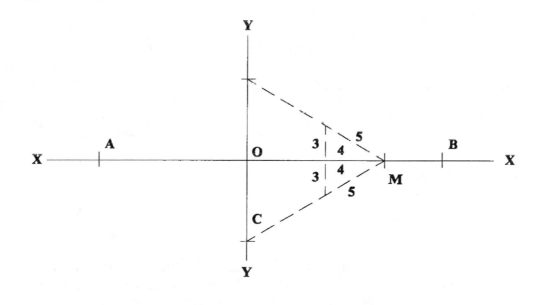

Figure 5-1 *Setting a perpendicular line through known points*

You want to draw a line (*YY*) through points *C* and *D* that is perpendicular to line *XX* at *O*. First, sketch the situation on paper to double-check how the figure will look. Then, use a tape to place line *YY* through *C* and *D*. Or, if the distance is too long for a tape, measure the line with a transit.

Using an Established Point

Assume your survey area is the one shown in Figure 5-1 and you need to make a line pass through point *O*.

Start by setting up the transit at point *O*. Then, line in on either point *A* or point *B* and flop the telescope to see if the line of sight is straight from point to point.

Next, turn 90° angles from points *A* and *B* to points *C* and *D*. Points *C* and *D* are arbitrary, put them wherever you like. Now if you draw a line through *C* and *D*, it perpendicularly bisects line *XX* and passes through *O*.

What if you cannot set up on point *O*? You can use a point anyplace on line *XX*, for instance, point *M*. You need to measure equal distances through equal angles from point *M* to both points *C* and *D*.

If you are using only a tape, use the 3-4-5 method to establish equal angles from point *M* on each side of line *XX*. (You can use any multiple of 3-4-5, say, 9-12-15.) If you use a transit, set up on any point on line *XX* and turn equal angles on each side of line *XX*. Measure equal distances along these angles to establish points such as *C* and *D*.

In order to pass the line through an established point, in this case point *O*, you must solve the triangles *MOC* and *MOD* to establish the arbitrary point *M*. Use the method for the 30°-60°-90° triangle described in Appendix B (Figure B-19).

Finding a Perpendicular Line

When you need to find a perpendicular to a line from a point that isn't on the line, use the method shown in Figure 5-2.

Start at point *O*, and drop a line perpendicular to line *XX*. Measure any selected distance, try 50 feet, from point *O* to a point that falls on line *XX* at *A*. Repeat these steps for line *OB*. Then, *OA* = *OB*, and *AC* = *BC* and a line from *O* to *C* is perpendicular to line *XX*.

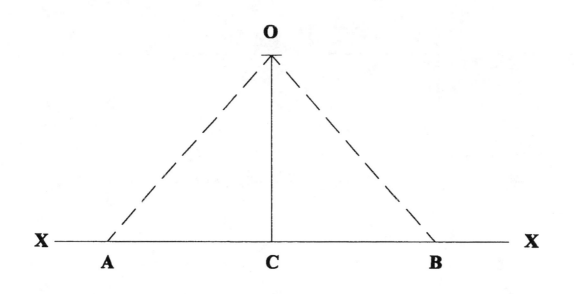

Figure 5-2 *Setting a perpendicular line from an unknown point*

You know that angle *AOC* equals angle *BOC*. To turn equal angles from point *O* with a transit, sight and mark the points *A* and *B* on line *XX* The halfway distance from *A* to *B* is point *C*.

Establishing Parallel Lines

To establish parallel lines, use the method shown in Figure 5-3. Assume the line *AB* exists and that line *CD* is to be parallel to *AB*.

Start by marking points *W* and *X*. You will find that the lines *WY* and *WZ* are equal and equal to the distance you want between the parallel lines.

If the lines are close together, you can measure with a tape. Stretch the tape from *X* to *W* to *Z* to *X*. Make *XW* 4 units, *WZ* 5 units, and *ZX* 3 units. This is a 3-4-5 triangle with a 90° angle at *X*.

Next, repeat this method for the triangle *WXY* and mark point *Y*. Now, if you put a line (*CD*) through points *Z* and *Y* it will be parallel to line *WX*.

If the lines are too far apart to use a tape, then use the transit. Simply set up at both points *W* and *X* and turn a 90° angle from line *AB* at both points to establish points *Y* and *Z* and measure equal distances from line *AB*.

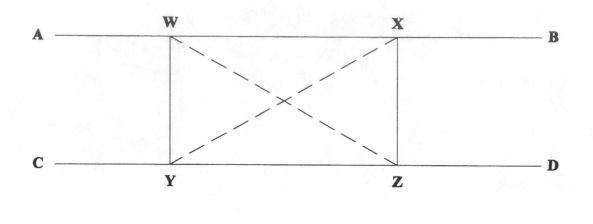

Figure 5-3 *Establishing parallel lines*

Dividing a Line Into Equal Parts

Suppose you need to divide a line into 11 equal parts as shown in Figure 5-4. Mathematically this is easy if you assume the line is 350 feet long. You just divide 350 by 11 to get 31.8181818. But how do you establish these equal sections in the field when the dimensions are critical to structures such as a line of column foundations?

Remember that field work and office work are directly related. So, start at your drawing board and lay out line *AB* as 3.5 inches long (1 inch equals 100 feet). Then, draw vertical lines at each end, at 90° to line *AB*.

Next, put the zero mark of an engineer's 10th scale on point A, and move the scale obliquely downward until you have a system of 11 equal divisions of the scale between point *A* and some point *C* on the right-hand line. Figure 5-4 shows a 30 scale used to make the divisions.

Now, extend the lines vertically from the division points on line *AC* to intersect line *AB* and establish the 11 equal divisions of line *AB*.

Use the same process with transit and tape, to lay out the line at the building site. For very accurate work, you will need to use both the transit and the tape to align and measure.

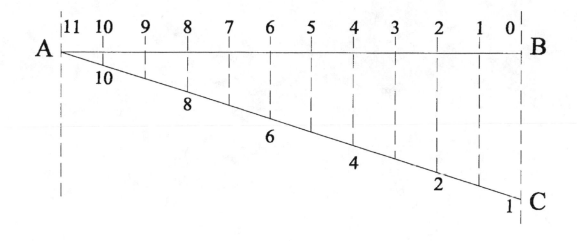

Figure 5-4 *Dividing a line into equal parts*

Drawing a Tangent Circle

The tangent point of two circles is on a line that connects their centers (see line *XX* and point *O* in Figure 5-5). The tangent point of a straight line (line *YY*) and a circle (either circle in Figure 5-5), is on the perpendicular (line *XO*) from the circle's center.

To draw a circle tangent to a given line and passing through a given point, you must make the circle center equidistant from both the line and the point. For instance, line *XX* in Figure 5-6 is parallel to the given line *MN* and the radius *R* distance away. If you use point *P* as center, and draw an arc of the given radius through line *XX*, the point of intersection, *O*, will be the center of the circle.

Similarly, in Figure 5-7, a circle is tangent to a line at a given point and passes through a second given point. Connect the two points P_1 and P_2 and draw the perpendicular bisector *MN*. Draw a perpendicular to the given line (*XX*) at point P_1. The point where this perpendicular intersects the line *MN* is the center *O* of the required circle.

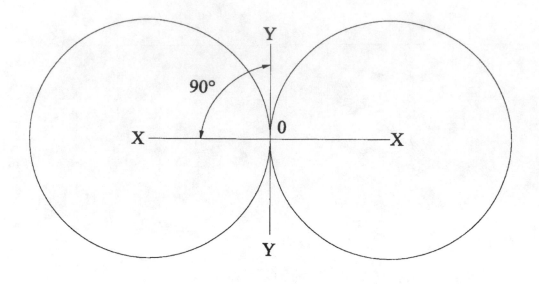

Figure 5-5 *Tangent point on the perpendicular*

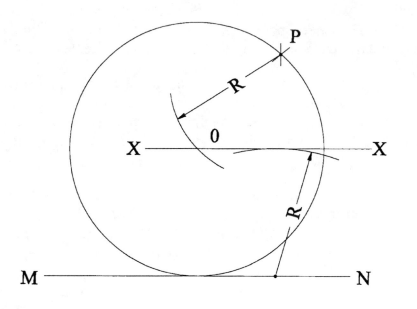

Figure 5-6 *Line XX, parallel to line MN*

Laying Out Angles

■ Using a Transit

Angles originate at some point in space. In Figure 5-8 the angles originate on line *XX* at point *O* and turn counterclockwise. To sight these angles in the field, set up the transit at point *O* and turn angles off the transit scale. Complementary angles are two angles with their sum equal to a right angle (90°) and supplementary angles are two angles with their sum equal to a straight angle (180°).

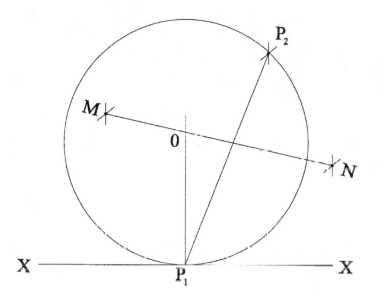

Figure 5-7 *A circle drawn tangent to a line*

To bisect an angle, set up on the vertex point and turn one half of the angle by scale.

To divide an angle, divide the angle by the number of parts you need and then turn these divisions from the vertex. For instance, six angles at 5° each is 30°/6 = 5° each. Thus, if you want to divide 45° into five equal parts: 45°/5 = 9. From the vertex, turn by scale from *O* to 9° to 18° to 27° to 36° to 45°.

Or, to construct a 45° angle, lay out two equal legs at 90° and join their ends with a (hypotenuse) line.

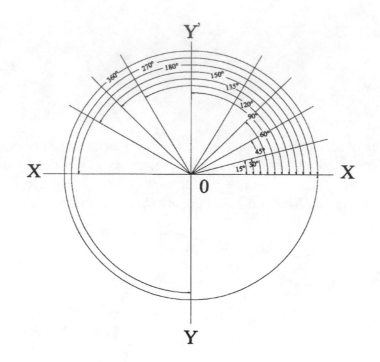

Figure 5-8 *Sighting angles*

■ Using a Tape

You can easily lay out an angle by tape measurement using the tangent method. For instance, Figure 5-9 shows an angle of 30° established from a point *O* on line *MN*. The tangent of 30° is expressed by tan 30° = *Y/X*.

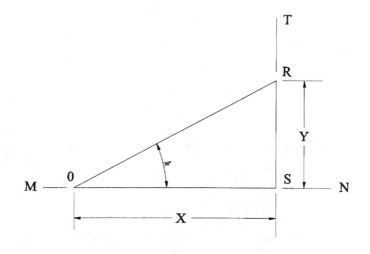

Figure 5-9 *Setting angles with a tape*

To find this, assume any convenient length for X, say 30 feet, and lay it out from O to S. Use a table of tangents to find the tangent of 30°, which is .5774. Then, tan 30° = Y/X; Y = tan 30° times X = .5774 times 30 = 17.322.

Next, measure 17.322 feet along line ST and at 90° to line MN. A line from O to R gives you the 30° angle. (If you use the transit to turn a single angle, set up at point O, sight in on line MN, and turn 30° by scale.)

Drawing a Circle Through Three Points

Only one circle can be drawn through three points. You use this fact to find the radius and center of an arc. The center of any arc lies on the perpendicular bisector of any chord of the arc.

For instance, Figure 5-10 has three points A-B-C through which a circle is drawn. Establish a line connecting AB and BC. Use points A and C to bisect the line connecting them. Then use points A, B, and C to bisect AB and BC. The bisectors of these lines intersect at a point O on the bisector of AC.

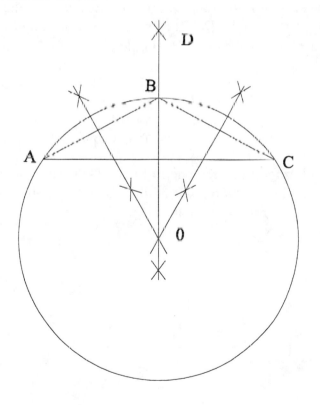

Figure 5-10 *Drawing a circle through three points*

This point O will be the center of a circle you draw through the points A, B, and C. Point O also is the center of the arc ABC and lies on the perpendicular bisector of the chord AC.

In the field, use two tapes, and measure equal distances. The size of the circle establishes the distance (it is the radius). Where the equal distances meet, set point O. In like manner, set point D. A line between these two points bisects the chord AC.

If the distances are too far to measure with a tape (or tapes), you can extend the lines by transit. Set stakes at given distances in the transit line of sight. And then turn the same degree of angle off both points A and C.

Drawing a Hexagon

Figure 5-11 shows a regular hexagon constructed with a given distance AD across the corners. To lay out this figure on paper establish point O and use AD as a diameter to draw a circle. Using points A and D as centers and the circle radius, draw arcs through center O and intersecting the circle. Connect all of the points with lines to form the hexagon.

In the field, use a transit to turn 60° angles off points A and D at the ends of the distance across the corners. Angles BAO and CDO, for example, are 60° angles. Measure along the line of transit sight and establish the sides. The circle is now divided into six arcs: 360°/6 = 60°.

If you are given the side lengths instead of a distance across the corners, it is always best to use the transit method. Simply turn four angles from a given line such as AD in Figure 5-11, measure off the four sides AB, CD, DE, and AF, and connect their endpoints.

If you already have a hexagon like the one in Figure 5-11, and you need to draw a circle on it, just remember that the intersection point of the perpendicular bisectors of any two sides is the circle center. This is true for any regular polygon.

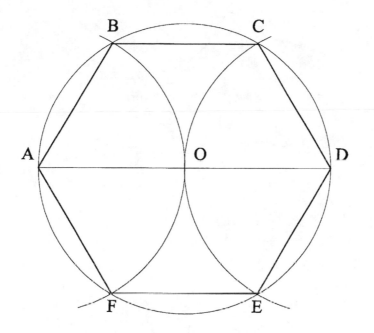

Figure 5-11 *Drawing a hexagon*

Ellipses

Occasionally, you will need to lay out an ellipse for a structure, flower bed, or patio.

■ Characteristics of an Ellipse

Figure 5-12 is an ellipse. It has two diameters at right angles. One is the major diameter, *XX'*, and the other is the minor diameter, *YY'*. The center is at *O*.

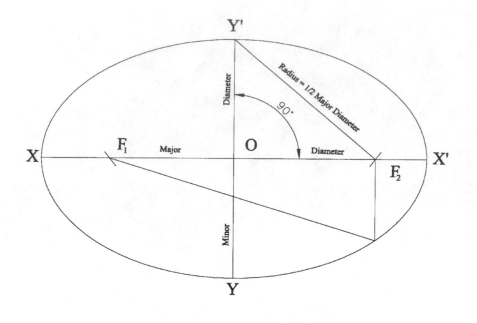

Figure 5-12 *Drawing an ellipse*

The diameter lengths determine the fatness or thinness of the ellipse. They are usually determined by whatever concept the designer has in mind. If the minor diameter is lengthened the ellipse almost becomes a circle. If the minor diameter is shortened the ellipse becomes as slender as a cigar.

The points F_1 and F_2 on the major diameter are called the focuses.

They can be located by striking the major distance with an arc that has a radius equal to one half of the major diameter and is centered at one end of the minor diameter. The distance F_1P plus the distance F_2P equals the major diameter. This distance determines the curve of the ellipse.

■ Drawing an Ellipse

Assume you need to lay out an ellipse with a major diameter of 30 feet and a minor diameter of 18.25 feet on a lawn to make a flower bed such as in Figure 5-12.

In the field, measure *XX'* at 30 feet in the desired direction and mark it by putting a surveyor's arrow into the soil at points *X* and *X'*. In the same manner measure and pin the points *Y* and *Y'* for the minor diameter at 90° to the major diameter.

Next, tie a loop in the end of a strong, non-stretch cord. Pin the endpoint of the loop at point Y'. Then form another loop in the cord so that with a surveyor's arrow in the loop, and the cord stretched, you can scratch a small arc in the soil at point F_2. The length of the cord from Y' to F_2 must be equal to 15 feet, one half of the major diameter.

In the same way, strike an arc at F_1. The two points F_1, and F_2 are the points of focus you will use to draw the ellipse curve.

Next, pin a strong, non-stretch cord at point F_1. Insert an arrow into a loop in the cord at the other end such that the length of the cord from F, to the arrow is equal to the length of the diameter. Pin this end at F_2. This length is also equal to the length $F_1P + F_2P$; and, it is equal to $F_1X' + X'F_2$ and $X'F_1 + XF_1$. Each of these sums is equal to the major diameter.

For convenience, use loops at each end. If you pull the slack in the cord into a loop and stretch it from F_1 to X', the cord will reach from F_1 to X' and back to F_2.

Place a surveyor's arrow in the loop of the cord at X'. Hold the cord taut with the arrow and move the arrow point along the earth to scratch out the curve of the ellipse as shown in Figure 5-12. As you hold the cord taut and mark the earth from X' to P, the cord will stretch from F_1 to point P and around the arrow to its anchoring point at F_2.

When the curve mark reaches X, the cord will twist with the long length over the short length. Now remove the arrow, straighten the cord, reinsert the arrow, and continue the curve back to the beginning point X'. You have marked a complete ellipse into the lawn.

Pentagons

You may be called upon to design a unique structure to draw the attention of shoppers to a shopping center. You might use the involute of a pentagon, a circle, or a spiral of Archimedes to make such a dramatic structure. You trace an involute as a spiral curve using a point on a taut cord as the cord unwinds from around a polygon or circle. The involute of a pentagon is drawn from the vertices of the pentagon.

■ Drawing a Pentagon

Figure 5-13 shows a pentagon inside a circle. You can size the pentagon by choosing a circle diameter that is slightly larger than the distance across vertices A and D of the pentagon.

To lay out the pentagon, draw perpendicular circle diameters. From points O and X' strike arcs to bisect OX' at M. Take the distance ME as a radius. Use M as a center to strike arc EN.

Next, use distance *EN* as a radius and *E* as the center to strike the arc *AN*. *AE* is one side of the pentagon. Now, use a compass set to distance *AE*, and step off the sides *AB*, *BC*, *CD*, and *DE* of the pentagon to draw them.

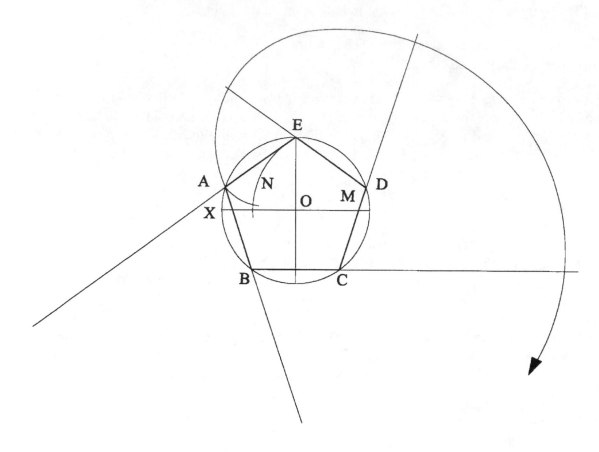

Figure 5-13 *Drawing a pentagon*

■ Drawing the Involute of the Pentagon

Now, draw the involute. Use point *E* as the center and side *AE* as the radius, and draw the curve from *A* to the extension of side *DE*. With point *D* as the center, and the distance from *D* to the point of intersection of the last curve drawn with the extension of side *DE*, extend the curve to the extension of side *CD*.

Finally, increase the radius by the length of one side for each curve. Continue around the pentagon vertices as much as you like. If the circle diameter were 5 feet, an involute curve from point *A* around the pentagon to the extension of side *AE* would be quite large.

The Involute of a Circle

Draw the involute of a circle as shown in Figure 5-14. Partially divide the circle into 30° angles. Draw a tangent to each radius. The tangent should be at 90° to the radius.

Next, set a compass at point 1. With a radius from point 1 to point Y, draw arc Y5. Then, set on point 2 and with radius 2-5 draw arc 5-6. Set up on point X and with radius X-6 draw arc 6-7.

Repeat the steps and continue the involute around the circle as far as practical.

■ Drawing the Spiral of Archimedes

A spiral of Archimedes is drawn as a number of divisions of a circle. In Figure 5-15 the outer line is the circle. It is divided into eight parts by using 45° angles. Assume you need to draw a radius to show each angle.

Start by drawing an arc with O as a center from number 1 on the vertical radius to the number 1, 45° radius. Then draw an arc from number 2 on the vertical radius to the number 2 radius.

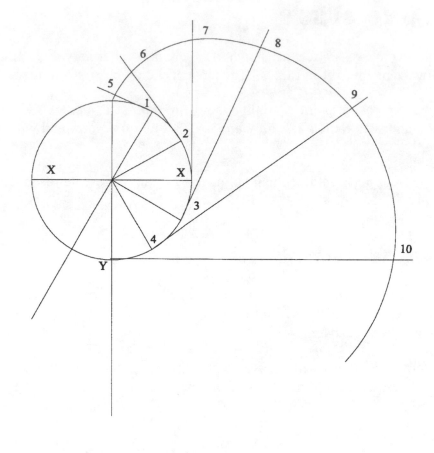

Figure 5-14 *Drawing the involute of a circle*

Repeat these steps for all numbers. Then, if you draw a smooth curve through the points where the arcs intersect a dividing radius, 0 to 1, 1 to 2, 2 to 3, and so on, the spiral will be complete for the given circle at point 8. Of course, if only the spiral is to be shown (or used for some purpose), the circle, the numbers, and the construction lines must be removed.

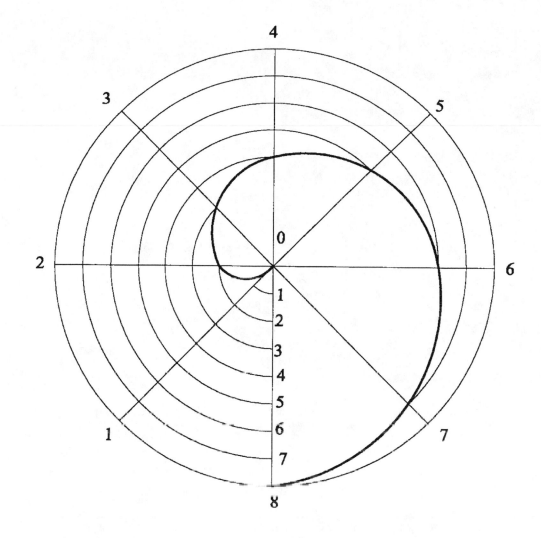

Figure 5-15 *A spiral of Archimedes*

CHAPTER 6

Stadia Surveying

Stadia surveying is a method of determining distances. A telescopic instrument with three horizontal cross hairs is focused on a graduated rod, and the distance is calculated from the difference of the projected top and bottom cross hair positions.

General Description of the Stadia Method

Stadia hairs (or *wires*) are horizontal in a transit or level, one above and one below the regular (centered) horizontal cross hair. Although it is not mathematically necessary for you to take a rod reading for all three cross hairs, doing so gives you valuable verification data.

The Tools

The term *stadia* refers not only to the method just described, but also to the rod and the wires you use. Stadia rods are available in a variety of sizes, colors, and gradations, but you can also use a standard leveling rod to take stadia readings.

■ Rods

Figure 6-1 shows a Lietz stadia rod divided into feet and tenths of feet. The solid diamond at the bottom is the 1-foot mark, the open diamond is the 1/2-foot mark, and the solid diamond at the top is the 2-foot mark. The numerals are red with black graduations on a white background and the diamond pattern is easy to read from far away.

You can find several other stadia rod face patterns to choose from. And, if the rod face becomes worn or damaged, you can replace it with a new face printed on special polyester material.

■ Reticules

Figure 6-2 shows a reticule that has extended stadia lines in a 1:100 ratio. When the top and bottom lines bridge exactly 1 foot on the stadia rod, the distance you are measuring is 100 feet.

You can buy Lietz reticules with either spiderweb or filaments or glass. Spiderweb cross-hairs are stretched across a cross-hair ring. One end of the spiderweb is secured on the ring by a drop of shellac. When the shellac hardens, the spiderweb is stretched tight and fastened in a similar manner.

Reticules of platinum or glass that are stretched across an annular ring, or a thin glass blank, which has filaments of metal deposited on lines ruled and etched in the glass, may be held in the ring.

Both types are accurate, but glass gives you greater durability and isn't affected by atmospheric conditions.

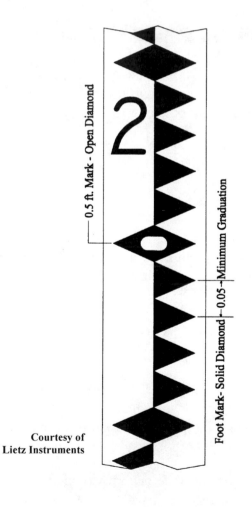

Courtesy of
Lietz Instruments

Figure 6-1 *A Lietz stadia rod*

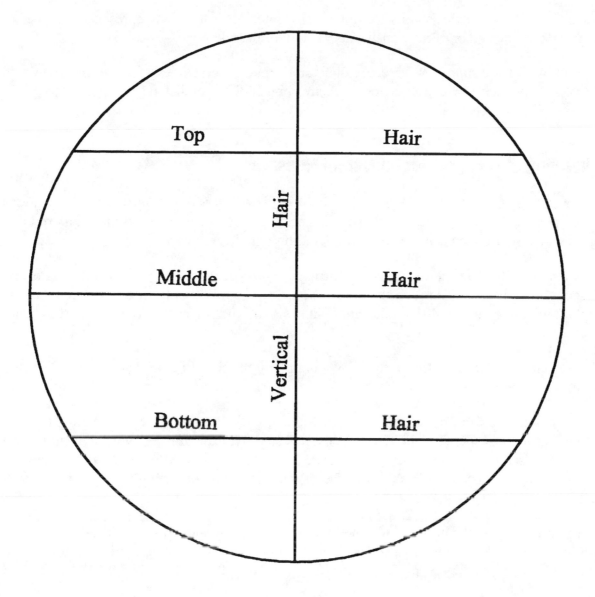

Figure 6-2 A reticule

Geometry You Will Need to Know

■ Similar Triangles

The principle of stadia is derived from the geometric principle that similar triangles have proportional sides. If you look through a telescope and see a 1-foot interval between the stadia hairs on a rod 100 feet away, and then look at a rod 200 feet away, you will see a 2-

foot interval between the hairs. Each 1-foot of interval on the rod represents 100 feet of distance from the center of the transit (the position of the telescope).

You will learn under "Correction Factors", that in practice, you also need to add telescope-dependent correction factors to the reading at different distances.

■ Deriving the Formula for Horizontal Measure

Figure 6-3 shows the geometric principle of stadia. Letters a and b indicate the positions of the upper and lower stadia wires. The distance A to B is the interval on the stadia rod that appears exactly between a and b when you look through the telescope.

A ray of light from point A passing through point F becomes parallel to the axis of the telescope after it passes through the lens at point a'. So, all rays from point A converge at point a' and are considered to be the single line $AFa'a$. (This fact is derived from the theory of lenses.)

The same condition is true for rays of light along the line $BFb'b$. The points where the oblique rays are changed to parallel rays are a' and b'.

The distance c from the center of the instrument to the center of the object glass (that is, the glass at the target end of the telescope) varies slightly, depending on your focusing. You must re-focus for each new position of the rod. When focused, $a'Fb'$ and AFB are similar triangles, and the distance $a'b'$ is to f as the distance AB is to s (generally shown as $a'b':f$ as $AB:S$).

The distance f is the altitude of triangle $a'Fb'$ and the distance S is the altitude of triangle AFB, because $S + (c + f)$ is the distance sought (from previously stated $a'b'$ is to f as AB is to S; also, $a'b'$ multiplied by S equals f multiplied by AB to yield $S = f$ divided by $a'b'$ multiplied by AB.

To calculate the distance, start by equating the ratio of sides for triangles $a'Fb'$ and AFB.

$a'b':f = AB:S$

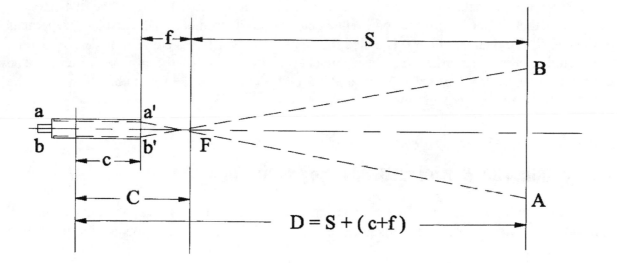

Figure 6-3 *The principle of stadia*

In fractional form you have:

$$\frac{a'b'}{f} = \frac{AB}{S}$$

Cross multiply to get:

$$a'b'S = f \times AB$$

Then,

$$S = f \times AB/a'b'$$

Notice that $a'b' = ab$, and is the interval between stadia wires. If you call this i, and call the rod reading R (distance AB), the equation is:

$$S = f/i \times R$$

The ratio f/i for any telescope is constant and the distance S is always equal to the rod reading multiplied by this f/i no matter where you position the rod.

■ Correction Factors

When you read a rod, think in terms of feet. Since 1 foot on the rod represents 100 feet of distance, then 1/2 foot on the rod represents 50 feet, 1/5 foot represents 20 feet, and so on.

You set up the instrument with its center over a given point. Therefore, the distance you measure is the distance from the rod to the center of the instrument. This distance, *D*, equals *S + (c + f)*. The distance *(c + f)* is called the *stadia constant*. This constant is determined by the manufacturer and marked on a card attached to the inside of the instrument case. The constant value ranges between 0.6 foot and 1.4 feet. You must add the stadia constant to the value *S* to get accurate calculations.

■ Calculating the Horizontal Distance on an Incline

So far, you have measured by stadia assuming the line of sight is level or nearly level. To measure by stadia when the line of sight is inclined, as in rough terrain, you need to make two corrections.

If the line of sight is up-slope (or down-slope), the distance measured is the same as measuring the inclined slope. You must correct this distance to a horizontal distance.

If the rod is held plumb, as usual, and is not perpendicular to the line of sight (the line of sight is inclined), then you see a longer portion of the rod between the stadia hairs than you would see if the line of sight were perpendicular to the rod. You must make a correction to this false reading. The corrections you make are based on formulae. The derivation of each formula is illustrated in Figure 6-4.

The line of sight is inclined at an angle θ (θ is the Greek letter theta) and you make an up-slope sighting. Holding the rod perpendicular to the line of sight gives you the stadia reading *a'b'* for the reading *FD* of the distance. When you add this to the constant *C*, you get the distance *BD*. To calculate the true horizontal distance *BE*, multiply *BD* by cos θ (trigonometric principle). The angle θ is read on the vertical vernier of the transit.

Holding the rod perpendicular to the line of sight is not practical, so the accepted practice is to hold the rod plumb. Because a plumb reading (*ab*) is longer than the perpendicular reading (*a'b'*), you must consider another correction.

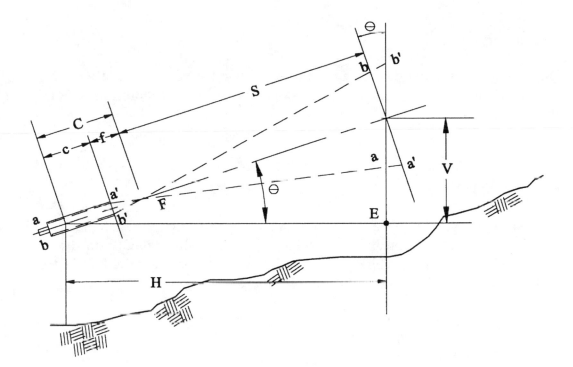

Figure 6-4 Calculating on an incline

If angle *bb'D* is almost a right angle, and angle *bDb'* is, by geometry, equal to θ, then ba × cos θ = a'b'. The *ab* distance you actually read is multiplied by cos θ to get the distance *FD*. Then,

$$BD = FD + C = f/i \times ab \cos \theta + C.$$

Let *R* represent the product of the stadia interval factor and the distance intercepted on the rod between the stadia hairs. Then, $R = f/I \times ab$, which, by substitution, gives you $BD = R \cos \theta + C$.

You calculate the horizontal distance *H* as follows. In the right triangle *DBE*, *BE* = *BD* cos θ. Then,

$$H = R \cos^2 \theta + C \cos \theta. \qquad \text{(Formula 1)}$$

or,

$$H = R - R \sin^2 \theta + C \cos \theta, \qquad \text{(Formula 2)}$$

where *H* = horizontal distance between the center of the instrument and rod, in feet; *R* = product of the stadia interval factor and the vertical distance intercepted on the vertical rod between stadia hairs, in feet; θ = the angle the line of sight makes with the horizontal; and C = the stadia constant, in feet.

The second equation is derived from the first using the trigonometric identity: $\cos^2 \theta + \sin^2 \theta = 1$. Either formula will apply to up-slope or down-slope sightings and both will give you the same results.

■ Example Readings

Horizontal Sighting. You have taken a stadia reading for which the stadia interval factor is 100.5 and the stadia constant is 1 foot. Find the horizontal distance from the transit to a vertical rod if the length intercepted on the rod between the stadia hairs is 7.27 feet and the angle that the line of sight makes with horizontal is 18°-21'.

Using formula 1,

$R = 7.27 \times 100 = 727$ feet
$\theta = 18°\text{-}21'$, and
$C = 1$ foot.

Then,

$H = R \cos^2 \theta + C \cos \theta$,

so,

$H = 727 \cos^2 18°\text{-}21' + 1 \cos 18°\text{-}21'$.

When you multiply and add,

$H = 727 \times .90088 + .94915$

$= 654.93976 + .94915 = 655.89$ feet.

Using formula 2,

$H = R - R \sin^2 \theta + C \cos \theta$,

$H = 727 - 727 \times .09911 + 1 \times .94915$

$= 727 - 72.05 + .94915 = 655.89$ feet

which is the same yield as formula 1.

In most instruments, stadia hairs are spaced so that $f/i = 100$. When 1 foot of the stadia rod is intercepted by the stadia hairs, the rod is $(100 + C)$ feet from the center of the instrument.

To test the stadia interval set the transit on level terrain with the plumb bob over a tack in a stake. This is one end of the test distance. Measure a 100-foot straight line (sight it in with the transit) and set the rod at the 100-foot distance. Have the assistant move the rod until the stadia hairs exactly intercept 1 foot on the rod. Assume that measured distance between the stake and the rod is 101.15 feet. This indicates that the constant C is 1.15 feet.

Move the rod near a 200-foot mark to a point where a 2-foot interval is intercepted between the stadia hairs. The distance measured should be 201.15 feet (not 202.30 feet). At 300, 400, and 500 feet you should observe 301.15, 401.15, and 501.15 feet, respectively.

When measuring with stadia you do not need to have one wire exactly sighted on an even-foot mark. Sighting on an even-foot mark only simplifies reading the rod. If you keep the telescope level, the stadia hairs will fall at random and you should read the interval with care.

Sighting Along an Incline. When the distance between two points is too far to read, assume an intermediate station about midpoint of the distance. Take a midpoint reading. Set up the transit at midpoint and take an endpoint reading. Add the two readings to get the total measurement.

With this information and the middle-hair reading, you can find the difference in elevation between the middle-hair reading and the center of the telescope:

$\sin \theta = DE/BD$, and

$DE = BD \sin \theta.$

Also,

$BD = R \cos \theta + C,$

and,

$DE = R \cos \theta + C \sin \theta.$

The difference in elevation is computed by the formula $V = R \times 1/2 \sin 2(\theta) + C \sin \theta$. The variables have the same meaning as in the foregoing horizontal distance problem.

Measuring the Vertical on an Incline Sighting. What is the difference in elevation between the ends of the sight in the previous example? Again, see Figure 6-4.

$R = 727'$, and $\theta = 18° - 21'$

So,

$$V = R \times 1/2 \sin^2 (\theta)$$

$$= 727 \times 1/2 \sin 36° - 42'$$

$$= 727 \times .29882 = 217.23 \text{ feet.}$$

Also,

$$C \sin \theta = 1 \times \sin 18° - 21' = 1 \times .31482 = .31482,$$

$$V = R \times 1/2 \sin^2 (\theta) + C \sin \theta$$

$$= 727 \times 1/2 \sin 36°-42' + 1 \times \sin \theta$$

$$= 727 \times .29882 + 1 \times .31482$$

$$= 217.24 + .31482 = 217.55 \text{ feet.}$$

Practical Considerations

■ Recording the Measurement

Using stadia is an excellent method for doing preliminary work. With one setup on a lot corner, you can locate everything on the lot by azimuth or angles from a lot line and take stadia-distance measures to map the area. For a large area, you might need several setups.

Handle notes according to the type of work you are doing. The headings are usually Line, Azimuth (or Angles), Stadia Distance, Vertical Distance, Elevation, and so on.

Read the instrument and rod values carefully. Be thorough. A bit of extra time and attention to the readings can prevent a return trip to the field.

Read all three wires and identify each reading in the notes as a top, middle, or bottom reading. If a wrong reading is made, the difference between the middle reading and the top or bottom reading will quickly reveal the mistake.

Read azimuths and angles twice. Read the rod twice. Then, list both readings. Many conditions can cause refraction (air disturbance), which will affect your readings, especially the rod readings.

Remember, read all three stadia hairs and add the constant $C = c + f$ to each reading. And, do not mistake the middle hair for a stadia hair.

Also, remember that the best work is done at 300 feet and under, although longer sights are often better than adding a list of shorter sights. A good check against readings is to take a half reading and a whole reading. The half should be exactly one half of the whole. If only a half reading can be taken, multiply it by two. To do this, the middle hair must be exactly halfway between the stadia hairs.

For inclined sights do not forget to record the vertical angle reading.

Check through your notes before you leave the setup.

■ Calibrating the Stadia Wire

Suppose a blank rod is set up at a distance of 100 feet, you take a sighting, and mark an interval of one foot on the rod corresponding to 100 feet of distance. Continue to test the rod by reading it at distances of 200, 300, and 400 feet. If the rod has been correctly marked, you should subdivide the intervals and paint them on the rod.

If the stadia wires are slightly in error, you need to multiply each reading by a constant to obtain the true reading.

Suppose that several observations at varying distances give you the reading 1.037. Then all distances would be too long. You can reduce the distance by dividing by 1.037. If the interval reads 0.9991, it is too short. Then divide all readings by 0.9991.

CHAPTER 7

Topographic Surveys

A *contour* is an imaginary line on the earth's surface that passes through and connects all points that have the same elevation. A standard map doesn't show any contours; it just shows you the relative positions of the features and structures on the earth's surface. When you add contour lines, both the relative position and the elevation are shown. Such a map is called a topographic map. In addition to contour lines, *topographic maps* use the symbols in Figure 7-1 to identify many features of the area. In this chapter you will learn how to create these maps by doing a *topographic survey*.

Illustrating a Depression

Figure 7-2 shows how contours can be used to map a depression inside a hilltop. Whenever a closed contour line encloses one or more lower lines, as in this illustration, you know you are dealing with a depression without an outlet.

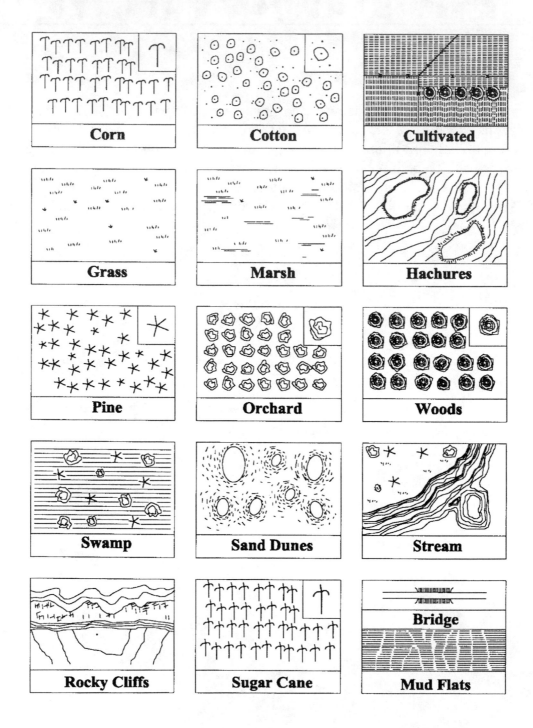

Figure 7-1 *Topographic symbols*

In Figure 7-2A, the hachure marks (facing outward) on the enclosing contour identify it as an ascending contour. The marks inside the three remaining contour lines identify them as descending contours.

In Figure 7-2B, the contours rise from elevation (El.) 120 to elevation 180, then descend to contour 140 inside a basin. The mark shown as El. 136 indicates the lowest point inside the basin. This basin has no outlet.

In Figure 7-2C, the hachure marks show that the slope rises through two elevations (contours), descends one elevation, then rises through two more elevations. The highest elevation is the mark shown as El. 204. This means there is a small hill inside the basin. The depth of water inside the basin would have to be at least 20 feet before it could spill over the two ascending contours.

Making a Topographic Survey

■ Contour Intervals

Since all of the points on a contour line lie at the same elevation, contour lines cannot cross each other (though they sometimes look as if they do when one is shown entering a cave or following the overhang of a cliff). You show successive contours using an elevation difference interval (*contour interval*) that meets your work requirements. The interval could be 1 foot, 20 feet (the interval used in most of the early government surveys), or any distance you choose.

Consider the purpose of your survey and determine an appropriate contour interval. For instance, if you are doing earthwork computations, choose a small interval of 2 feet.

■ Reference Lines

Generally, the topographic surveys you will do for building purposes will involve small areas of land. It will help you to use reference lines for your survey. A reference line could be a line that follows a property line, an observable line that extends along an entire side of the property, or a line that runs through the property.

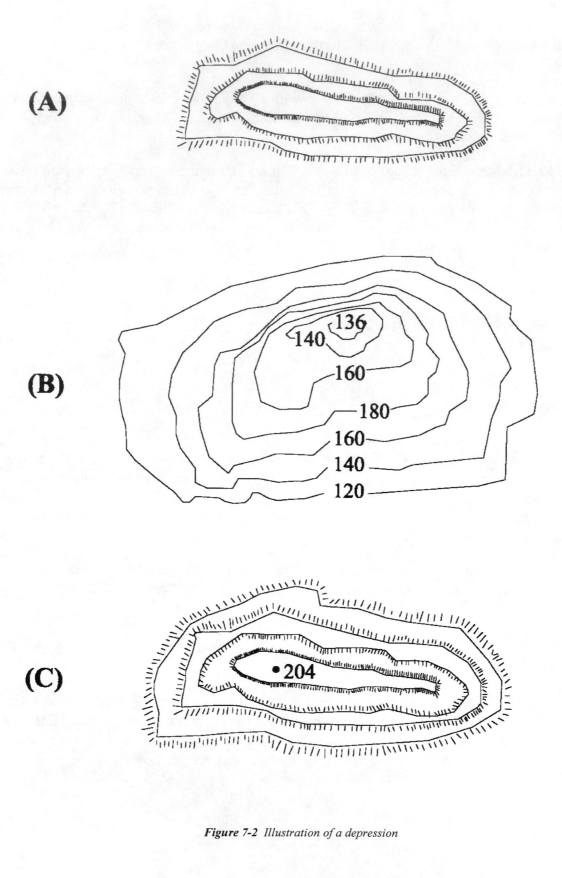

(A)

(B)

(C)

Figure 7-2 *Illustration of a depression*

If all of the important features of the area can be sighted and measured from property lines, then use the property lines as your reference lines. If you can reach all of the features except for those near the center of the area, perhaps you can map the center area from a road running through the area. If there isn't a road, put a reference line of stakes through the area.

After choosing your reference lines, continue your survey with a preliminary examination of the area. Decide how to organize your field work. Mark the reference lines and stations. Locate boundary lines, streams, utilities, buildings, roads, and other details.

■ The Direct-Contour Method

A highly favored method for locating contours is the *direct-contour method*. Choose a point, perhaps a property corner, for reference and establish a station point by azimuth from the property line and measuring from the reference point to the station point.

From this first station, set another station by azimuth off a line from the reference point to the first station and a measurement from the first station (station 1) to the second (station 2). You can establish all of the stations before you make contour sightings, or choose each station as the work progresses.

Sample Survey 1. Figure 7-3 illustrates the start of a topographic survey for a building site. Assume that this is your survey and remember that you will need to map the information, so your notes and field drawings must be accurate.

First, assume that El. 100.00 is relative. You will use a relative elevation only for the "job-at-hand"; it has no relationship to the actual geoelevation. Set a reference point with El. 100.00 at D. Next, establish the contour line that crosses the line AC at D as El. 100.00. From this reference point find the elevation of a point near or at the property corner A for a future reference point. (Note that portions of the boundary lines are shown as AB and AC.) 100

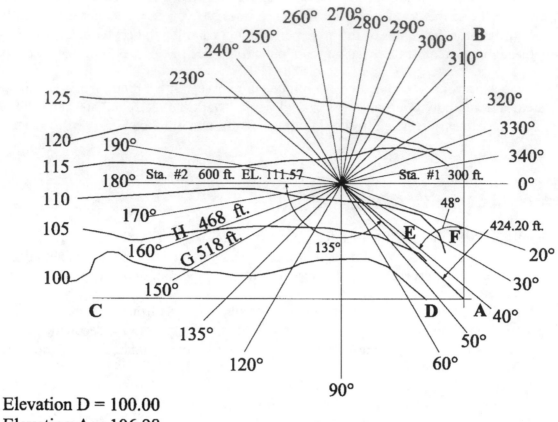

Elevation D = 100.00
Elevation A = 106.98

Figure 7-3 A topographic survey

Second, set up on *A*. Take a sighting along the line *AB*, and turn a 45° angle to establish station 1. Assume the station is 300 feet from both property lines. The distance from *A* to the station is then 300 × 1.414 = 424.20 feet. Limit your sightings to 500 feet or less, depending on the terrain and the condition of the atmosphere on the work day. Thus, station 1 is at El. 111.57 and El. 113.62 is found on the property line *AB* at 0°.

Third, set up on station 1. Take a sighting on point *A* and turn a 135° angle to the right. Measure a 600-foot distance along this line of sight to establish station 2. You will establish other stations from stations 1 and 2 as the work progresses. Now that you have set reference point *D*, 101 set up the instrument on station 1, and fixed the sight on 0°; you are ready to begin the actual surveying.

To take linear measurements you can use stadia surveying (see Chapter 6). Take elevations by direct rod readings. For this type of work you can get leveling rods of 25 feet or longer.

Consider the situation in Figure 7-3. Find the height of instrument (HI) on station 1 at El. 115.77. Turn a 20° angle left from 0° and take readings along the line of sight to the property line. All the readings are between El. 110.00 and El. 115.00.

If you make the turn 30° right, you will find two readings along this line of sight, one at point *E* and one at point *F*. Contour 110.00 crosses the line of sight at these points.

When you find a reading that corresponds to a contour line, take a stadia reading for distance. When you have both the distance measurements and the elevation measurements, you have plotted the area.

A reading at *G* gives you 518 feet and one at *H* gives you 468 feet. You will find only one reading, on El. 110.00, along the 1700 line of sight within or near 500 feet.

Since the *HI* is El. 115.77, before you move the instrument from station 1, search for El. 115.00. You will find it on sightings along all angles from 190° to 340°. Take whichever sightings will allow accurate plotting.

Make your next setup on station 2 and take the elevations 100.00, 105.00, and 110.00 for the area within 500 feet. Again, check to see if any of El. 115.00 can be established from here.

Follow this procedure until you have covered the area along which these elevations lie. Then you must establish stations above El. 125.00, and even higher, depending on the terrain and the extent of the work.

Sample Survey 2. In Figure 7-4 a rectangular (2000 × 3000 feet), 137.74-acre area of land needs to be studied for subdivision (2000 × 3000/43,560 = 137.74). The area used to be a family dairy farm, as you can see by the farmhouse, barn, silos, outbuildings, and orchard. As the farming operation was reduced, five areas adjacent to the lake were fenced and recreational cottages were built on them for summer rental.

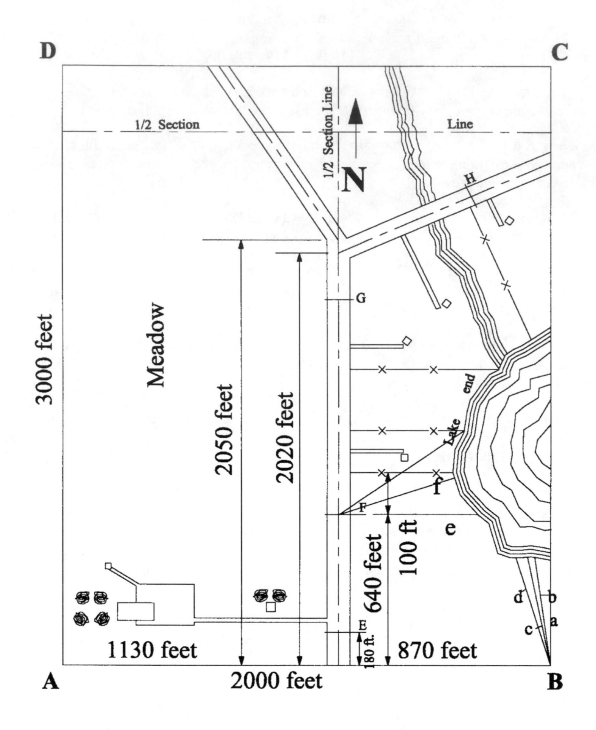

Figure 7-4 Land to be subdivided

Consider the topographic survey you would make to study the potential for subdivision. The sightings are limited to 600 feet. Establish your stations at the southeast property corner (B), and on the section line running along the road centerline (E and F). (You can locate stations at any convenient point.)

From station B, take sightings to establish the lake's shoreline. Read the angles to the nearest minute and read distances to the nearest tenth of a foot. These are the turned angles (L = left and R = right) and the bearings you should use. All of the angles are turned from due north.

Sighting	Angle	Bearing	Distance
a	none	Due North	454.50 feet
b	L7°20'	N7°20'W	492.70 feet
c	L11°22'	N11°22'W	500.10 feet
d	L18°10'	N18°10'W	620.20 feet

Assume that the farmstead is to be demolished. Establish station E to find the location of the two trees near the house. These, and the orchard, are to remain. You can find more of the lake's shoreline from station F. Turn the angles from the road's centerline, looking north. Drive a spike at E, F, G, and on the centerline at the fork in the road junction.

Sighting	Angle	Bearing	Distance
e	90°	Due East	570.00 feet
f	R77°48'	N77°48'E	472.00 feet
g	R47°28'	N47°28'E	422.70 feet

Take sightings from all the stations, including H, along the road to complete the shoreline and creek-bank sightings (except for a short section of creek near the north property line). As the work moves northward, locate and include the cottages and fence lines on the lake shore.

To find the surface elevation of the lake, dig a small hole near the waterline. It will fill with water to the lake's surface elevation. Wind won't disturb it, as it does the lake's surface, and you can take a sighting at the water level in the hole.

You may use numbers or letters for the points of reference. And you record angles by direction, L or R, and list the bearing. Be sure to sketch the work with adequate detail and in a way that's easy to understand. Do not leave any threads dangling. Gather all the information in one trip or a series of trips and then map it.

CHAPTER 8

Leveling

For surveys that involve measuring the elevations of buildings, roads, or grading you will often use a level. This chapter describes the types of levels available and their uses. Then, the two following chapters take you through a variety of simple to complex leveling situations, including how to measure cut-and-fill areas for grading.

Leveling Instruments

Both the Wye level and the Dumpy level are widely used by builders, contractors, and engineers. On the Wye level, the spirit level (the part with the glass tube with a bubble, usually filled with alcohol to prevent freezing) is attached to a telescope tube, which is clamped to two Y- shaped bearings; hence its name Wye level. The Dumpy level is made in one casting and has few movable parts to get out of adjustment.

Figure 8-1 shows the Lietz Engineers Level 180. Its supports are bronze, soldered to the telescope, and screwed to the level bar for extra rigidity. The long telescope gives you great accuracy. The Lietz instrument is built for preciseness and it is rugged for long, useful service. In fact, any leveling instrument handled with care will last many years. Some engineers have used their levels for more than 50 years.

■ Setting Up the Level

For leveling, set the instrument up so that you can make as many sightings as possible from one location. Spending less time in the field leaves you more time in the office to do your computations. Set the instrument in a safe place. If construction is going on and men and machinery are moving around, your instrument could be damaged.

Place the tripod legs firmly and simultaneously bring the instrument as nearly level as you can by eye. Position the telescope over two leveling screws and bring the bubble to the center of the tube. Then turn the telescope into position over the other two screws and bring the bubble to the center of the tube. Repeat this leveling process from position to position until the bubble is centered at each position. After this, the bubble should remain centered during an entire horizontal revolution of the telescope.

Figure 8-1 *The Lietz Engineers Level 180*

■ Reading Levels

You use the level to take a reading on a leveling rod held in a plumb (vertical) position by an assistant. You can also get a rod level, especially designed to fit against a rod edge, to keep the rod plumb.

Figure 8-2 shows a partial view of a Philadelphia leveling rod face. The rod readings show the fine degree of accuracy you can get. The readings begin at the bottom of the rod, marked zero, and proceed upward. The first reading shown is short of 1 foot, and is therefore a decimal reading of 0.85.

Study the readings. They show you how to read a rod that is graduated in feet and tenths of a foot. Rod faces are available in feet, tenths, hundredths, meters, decimeters, and centimeters. Also, you can get a rod marked for direct elevation work or one marked for stadia readings (see Chapter 6). The rod in Figure 8-2 extends to 13 feet. The Lietz telescoping fiberglass rod reads to 25 feet.

You can also get targets for these rods to make accuracy a certainty. In the target reading in Figure 8-2b the number of feet is given by the number immediately above the three, in this case it is a one. The three shows the number of tenths since it is below the zero of the vernier. The number of hundredths are counted from the three up to and including the tenth below the zero on the vernier scale. It is 0.03 feet.

The number of thousandths are shown by the vernier graduation that coincides with a rod graduation of six. The rod reading is 1.336 feet.

Periodically, you should test your rod for accuracy. If the rod is too worn or is damaged by folding it up too roughly, your readings could be inaccurate. To test the rod, simply extend it and measure between the top and bottom markings with a tape.

■ Using Hand Levels

You can use a hand level in many situations such as backfill, preliminary grade sightings, height of small hills, and so on. For instance, assume you are doing a backfill sighting for a new residence. There should be a slight grade to drain rainfall from the future building.

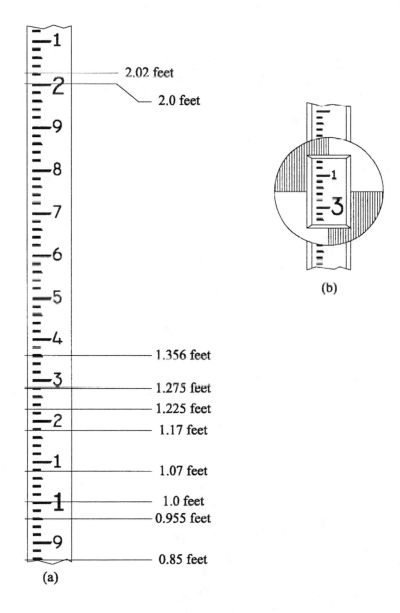

Figure 8-2 *A leveling rod*

Stand on a spot near the building. Bring the level to your eye and move it up and down until the cross hair exactly bisects the bubble. Place a small mark on the building at the point of sighting. Move 5 feet from the building and sight on the mark. If the grade recedes from the building, the second level sighting will be lower than the first.

Sewer grades should not exceed a slope of 10 feet per hundred feet of sewer line. You can use a hand level to quickly determine the natural slope of a hill and whether a drop manhole is required. Measure downhill while standing on a spot at the top of the hill. The assistant stands at a spot down the hill so that a reading can be taken near the top of the rod. Take and record a reading.

Next, move downhill to the exact spot the rod was held for the first reading. The assistant should move downhill to provide a new sighting. Record the second reading and total the two readings.

Using a tape, measure the length of the slope along the line of sightings. If the drop in grade is more than one tenth of the length, use a drop manhole to reduce the speed of sewage flow to an acceptable limit.

CHAPTER 9

Basic Leveling Surveys

After setting up your level, you need to determine the plane of sight of the instrument by measuring up to it from a predetermined plane for which you already know the elevation.

The known elevation plane can be that of a monument (bench mark) placed by the U.S. Geological Survey or a relative elevation established on some point for temporary use during layout and construction activity. Generally, for building, you will use the latter. The relative elevation starts at elevation 100.00 (El. 100.00) and the temporary monument can be concrete, a long pipe driven into the ground, a nearby bridge pier, a step on a nearby building, or anything that is permanent for the duration of the leveling work. Once you have established the elevation for the line of sight, you can find the elevation of any other point by measuring down from the established line of sight.

Finding Elevations

Figure 9-1 shows the profile of a building site. The details of this site are given in Figure 9-2. Note how the entries are organized. In the column under station (Sta.), the bench mark (BM) is relative at elevation 100.00 and is located on a plus (+) mark chiseled into the top of a concrete culvert (see the "Remarks" column) 30 feet south of station 0.

Set up the level at any convenient place that gives you a clear sighting on both the bench mark location and station 0. The backsight (BS) on the benchmark reads 5.30. Add this to the relative elevation of 100.00 to get a height-of-instrument (HI) reading of 105.30. A foresight (FS) on station 0 gives a reading of 4.30. Subtract this from 105.30 (HI) to get the elevation of station 0 of 101.00.

In this situation, station 1 can also be seen from the present instrument location for a reading of 11.51 (FS). Subtract this from the HI of 105.30 to get the elevation of station 1 of 93.79.

Since the leveling rod is only 13 feet long, the level must be moved to a location from which both stations 1 and 2 can be seen.

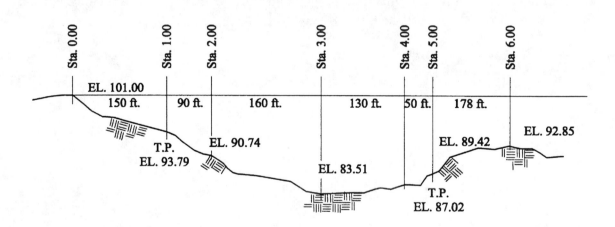

Figure 9-1 The profile of a building site

Profile for Building Lot No. 104, this city. Notes in Field Book 10, 2000.					July 28, 2000 D. Exline - Level J. Murphy - Rod		
Sta.	B S	H I	F S		El.	Remarks	
BM	5.30	105.30			100.00	+ on culvert 30' south of Sta. 0.00	
0			4.3		101.00	The readings were taken on a straight line at the distances shown on the drawing herewith. (Fig. 9-1)	
1	1.32	95.11	11.51		93.79		
2			4.37		90.74		
3	8.54	92.05	11.60		83.51		
4			5.03		87.02		
5	7.36	96.78	2.63		89.42		
6			3.93		92.85		

B.S.
22.52   F.S. on T.P.
+ 100.00   25.74
122.52
- 25.74
96.78

Figure 9-2 Site details

From the new location, take a backsight on station 1 (El. 93.79). The BS of 1.32 plus the El. 93.79 gives you a new HI of 95.11. A foresight on station 2 of 4.37 is subtracted from the HI (95.11 −4.37) to give you an elevation of 90.74 for station 2. A foresight on station 3 of 11.60 gives you an elevation for station 3 of 83.71 (HI minus FS is 95.11 − 11.60 = 83.61).

Again, you will need to move the instrument to a new location and level it. A backsight on station 3 reads 8.54 for an HI of 92.05 (83.51 + 8.54 = 92.05). A foresight on station 4 reads 5.03. Subtract this from 92.05 to get an elevation on station 4 of 87.02. A foresight on station 5 reads 2.63. Subtracting, 92.05 − 2.63 = 89.42, gives you the elevation of station 5.

The instrument needs to be moved again. A BS on station 5 reads 7.36. Adding this to the elevation of station 5 (7.36 + 89.42) gives you an HI of 96.78. A foresight on station 6 reads 3.93. Subtract this from the HI (96.78 − 3.93) to get 92.85, the elevation of station 6.

To check the level computations, add the sum of the backsights to the elevation of the starting point (22.52 + 100.00 = 122.52). From this sum subtract the sum of all the foresights made on turning points.

The turning points are stations 1, 3, and 5. The sum of the foresights is 11.51 + 11.60 + 2.63 = 25.74.

If your arithmetic is correct, when you subtract the sum of the foresights from the sum of the backsights and BM elevation, you will get the last HI or, 122.52 − 25.74 = 96.78, which is correct.

■ Keeping Notes

You can keep level notes in several ways. The chart method used in Figure 9-2 is very simple and concise. Notice that a blank column is provided. You can use this column to keep foresights where the turning points are not on a station, and where readings on stations are called intermediate stations (IS). Use the blank column for the IS and the FS column for foresights on the turning points.

Types of Leveling

You may have heard the term *differential leveling*. This simply means the difference in elevation between two points. When the two points are far apart, and perhaps have up and down grades between them, you take a line of levels (sightings on the rod) and record them using the arithmetic as given for Figure 9-2.

Figures 9-1 and 9-2 show *profile* leveling used to find the elevations of points a known distance apart in order to "profile" the surface along a given line. In Figure 9-1, the line of sighting and the elevation stations were chosen to give a profile of the area portraying as nearly as possible an accurate "lay" of the working area.

The term *flying levels* describes rapid and less precise work done to determine only approximate elevations, generally for a preliminary study of the area.

Cross-levels are elevations taken at right angles on either side of the main line of levels. For example, levels you might take along the centerline of a proposed highway with cross-levels along either right-of-way line.

Reciprocal leveling is the process of finding the difference of elevation between two points by taking two sets of levels. Set up the level near one point and take sights on both points. Next, set up the level near the other point and take sights on both points.

The longer and shorter distances should be about the same for both sets of sightings. Take the mean (average) of the two sightings as the true difference of elevation. This process corrects for curvature of the earth and refraction of the air and, in general, eliminates the error inherent in adjustments of the level.

For example, in Figure 9-3, *A* represents one point and *A'* represents another point. Points *B* and *B'* represent the instrument setups. The short distance is a (about 30 to 40 feet) and the long distance is *b* (whatever length it is). The true distance between the points is *d*. Distance is not important as long as you make clear sightings.

Average the sights from both *B* and *B'* to *A*. Average the sights from both *B* and *B'* to point *A'*. The mean of the two results obtained by this method is the true difference of elevation.

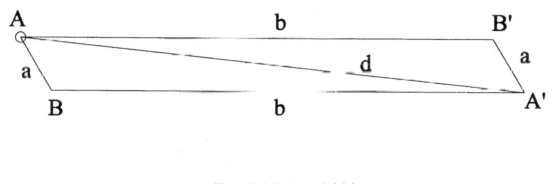

Figure 9-3 Reciprocal sightings

Grade Stakes

In raising buildings it is required that you set points at given elevations and set grade stakes. A series of points set below a datum line is shown in Figure 9-4A. The leveling instrument is set up to sight all five points from one location.

The term *grade* describes the degree of vertical rise or fall (inclination) for each 100 feet you measure horizontally. In this sense, grade means the *gradient of slope*.

A *level grade* has no rise or fall. A vertical drop of I foot in 100 feet horizontally is a 1 percent grade. A vertical drop of 2 feet in 100 feet horizontally is a 2 percent grade. A 5 percent grade is considered ideal for streets and roads as a maximum grade.

In Figure 9-4A, notice the distance below the datum line and the distances the points are to be separated. The point at zero is 4.3 feet below El. 100.00. Then, 120 feet away, point 1 is set 0.6 foot below El. 100.00. After you set point 1, turn the level 180° horizontally to sight on the other points. The level bubble must indicate a level instrument after the 180° swing of the instrument. When doing this work it is important that you have the instrument exactly level before you begin.

Use the first letter of the words grade, difference, and length in a formula: G = grade or gradient, D = difference in elevation between two points, and L = length horizontally between two points.

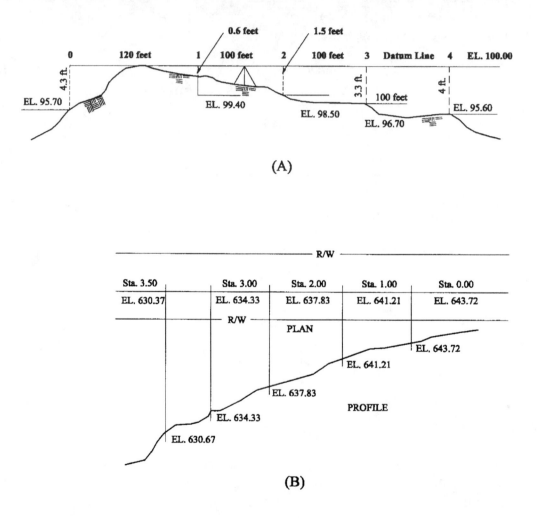

Figure 9-4 *Setting grade stakes*

Roadways have stations along the centerline 100 feet apart. Thus, it is easy to compute the grades. Figure 9-4B shows 350 feet of centerline at the beginning of a roadway, station 0.00 through station 3.50. The centerline arrow points down grade. A profile of the elevations also is shown.

The grade between station 0.00 and station 1.00 is $G = D/L$. $G = (643.72 - 641.2)/100 =$ 2.51/100 = 0.0251, about 2.5 percent. The grade between station 1.00 and station 2.00 is $G =$ $(641.21 - 637.83)/100 = 3.38/100 = 0.0338$ percent. Between station 2.00 and station 3.00 the grade is 0.035 percent. Between station 3.00 and station 3.50 the grade is $(634.33 - 630.67)\ 0.5/100 = 0.0183$ percent.

The distance is one half of the station distance, 50 feet. That's how you get the 0.5 multiple.

The grade of the finished roadbed is an average of the grades from station 0.00 through a given station along the centerline, depending on the "lay of the land" ahead. For example, if the terrain nearly levels out at station 3.50, then (El. 643.72 – El. 630.67)/350 = 0.03728, about a 0.0373 percent grade. This means the road is at 3.73 feet at each 100 foot station. This is a calculated rod reading. Elevations will be set on each centerline station at a drop of 3.73 feet and constructed to this elevation by cut or fill.

Elevations will also be taken on each right-of-way line at right angles from each centerline station to establish the existing grade there. A profile is worked out for each right-of-way line.

Potential for Error

Errors in leveling come from many sources. You should test new levels and leveling rods thoroughly. And, be sure to periodically test them for wear and damage. Also, when you travel over rough terrain, anchor the equipment firmly in the vehicle so that it doesn't move around.

When you set up the level, avoid mistakes such as not centering the bubble, over-clamping the spindle, failing to watch the bubble, leaning on the tripod or telescope, stepping heavily near the tripod legs, and bumping the transit. Other errors are not having the leveling rod plumb, having soil on the base of the rod, mistakes in rod readings, and errors in sighting. Another factor is personal error. Some people are clumsy, use poor judgment taking far sightings by sighting too low or too high, and practice alternately taking a short sight and a long sight. Make your sightings as nearly equal in distance as possible.

If your backsights and foresights are greatly unequal, use reciprocal leveling. And, if the leveling is truly critical, rerun the leveling line in the opposite direction. This is a good check.

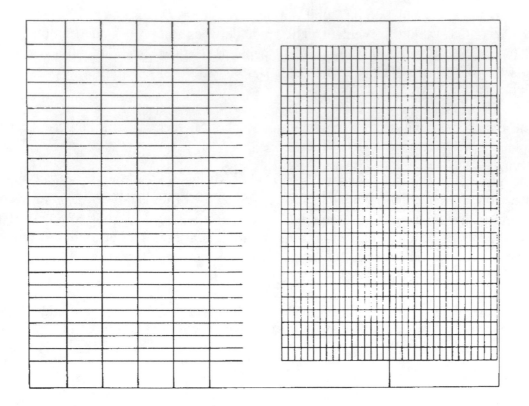

Figure 9-5 Sample field book pages

Level building foundations and piers from setups at two or more locations.

Remember also that high winds and extreme temperatures can cause errors. If you must do leveling under these conditions, put up protective shields to block the wind. Also, use an umbrella or other shelter to shade the instrument to combat extreme temperatures.

Finally, every morning check two or three bench marks or turning points for tampering or changes due to rain, frost, or other natural causes.

If you give careful attention to avoiding all of the above problem areas, you can produce almost error-free results in your work.

Figure 9-5 shows sample facing pages from a surveyor's field book. Keep level notes and values on the left-hand page. Keep a field drawing as the work progresses on the right-hand page. Since you can keep notes in a number of ways, the columns are left blank. You choose the column headings you prefer.

CHAPTER 10

Complex Leveling

Now that you have worked through some simple leveling situations, this chapter examines some more complicated situations involving streets, curves, and grading.

Evaluating a Site

As a rule of thumb, any building site large enough to include streets must be carefully studied. Use a field study to decide whether the area suits your needs. Then, if it seems suitable, do a reconnaissance survey (rapid but thorough examination of the entire area, especially contours) to decide on the possible locations for building sites and utilities.

After you have located all possible building sites, make another study to fit the streets into the plan. Deliberate, thoughtful handling of the details gained from this survey can save a lot of time and expense.

Planning the Streets

If the site is for residential housing, give streets close attention. Among the residents will be poor drivers, newly licensed drivers, and student drivers, so do not design hard-to-follow streets with dangerous intersections. And, unless the area is truly mountainous, keep the street grades below 5 percent.

■ A Simple Plan

At this stage of planning, a profile of the centerline of proposed streets is all you need. You can add curb lines and right-of-way lines later.

For example, suppose you are laying out street plans for a slightly hilly area. Using the letters D, G, and L in an equation, you can lay out a street to a given grade (say, 5 percent) on a contour map of the site. Such a layout is shown in Figure 10-1. The centerline of the 5 percent grade is established by connecting the points m, n, o, p, q, r, and s on contours with 5-foot intervals. Thus, for a grade that will not exceed 5 percent on a contour interval of 5 feet, the length $L = D/G$. $L = 5/0.05 = 100$ feet, the minimum length of street between any two contour lines.

Beginning with the view in Figure 10-lA, choose a point such as m, as a center, and strike an arc with a radius of 100 feet to intersect contour 735 at point n. Then, using point n as a center and a 100-foot radius, find point O on contour 740. Use the same method to find all other points.

Notice that this route creates a condition on contours 735 and 740 in which the street follows each of these contours for a distance. Since the street is level along the sections n to t and o to u, all of the grade falls in the sections t to o and u to p. This means that the grade will exceed the 5 percent you want.

The view in Figure 10-1B shows a half-contour interval, 735.5, drawn between contours 735 and 740. A one-half radius (100/2 = 50) is used to lay out arcs *nz* and *zy*. The distance along the combined lengths of arc is longer than 100 feet and the grade will be slightly less than 5 percent.

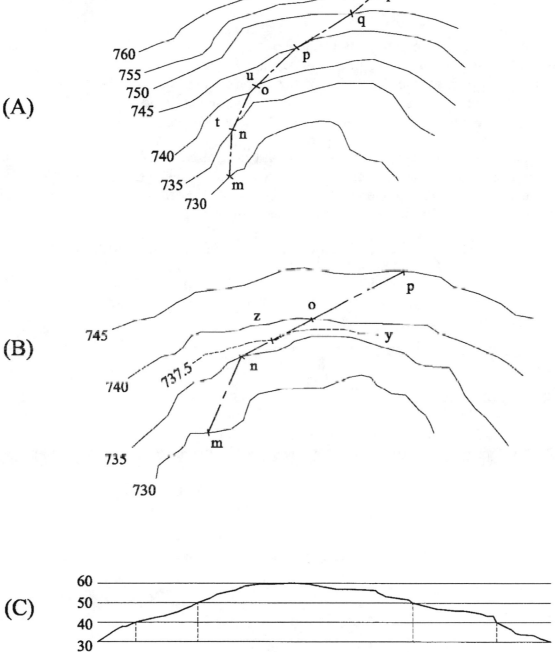

Figure 10-1 A simple street plan

Plan streets to keep cut-and-fill at a minimum while generating enough cut for a volume of soil to fill the low areas as much as you need.

■ Planning the Curb

After you have set the limits of the street and the cut-and-fill is completed, set grade stakes for the curb.

Line in the stakes along the curb line at about 25-foot intervals. Drive each stake a few inches into the soil. Place these intermediate stakes at grade for easy form setting. If the grade is to be 3 percent, the rise or fall of the curb is 3 feet for each 100 feet of length of street. At 25-foot intervals for grade stakes on a descending grade, the grade rod will read $3/4 = 0.75$ foot less for each succeeding stake as the grade descends.

In Figure 10-2 the stakes are shown along 400 feet of curb line length. The stakes are numbered 1 through 17 and the elevations are given in the table. The total drop in elevation for 400 feet with 3 percent grade is 12 feet. The instrument setup should be about midway between the ends of maximum sighting for the instrument height to allow one setup.

Mark a grade stake "Grade" and a cut-or-fill stake "Cut_____feet" or "Fill_____feet."

After the curb has been installed, mark the grade for the pavement along the curb. Make these marks at 25-foot intervals with a chisel. Put a row of stakes along the centerline to establish the crown of the street.

■ Planning the Sewers

Sewers and drainage pipes are generally laid on grades from a 0.2-foot drop per 100 lineal feet to a 10-foot drop per 100 lineal feet. The smaller the grade, the more accurately you should lay the sewer.

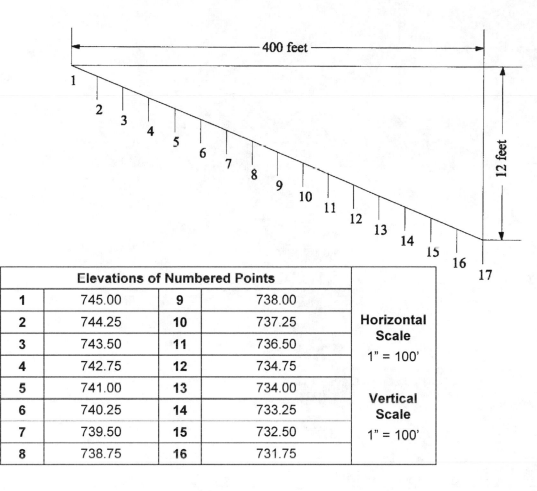

Elevations of Numbered Points				
1	745.00	**9**	738.00	
2	744.25	**10**	737.25	**Horizontal Scale** **1" = 100'**
3	743.50	**11**	736.50	
4	742.75	**12**	734.75	
5	741.00	**13**	734.00	**Vertical Scale**
6	740.25	**14**	733.25	
7	739.50	**15**	732.50	**1" = 100'**
8	738.75	**16**	731.75	

Figure 10-2 Setting stakes for a curb

To control the grade, establish the centerline of the sewer above the trench and set grades on it for the convenience of the pipe layers. Then, when excavation of the pipe trench is well underway, work on the elevated centerline can begin.

Figure 10-3A shows a structure with a cross-member above the trench strongly supported on stakes firmly set on each side of the trench. A 2-inch-wide strip (grade slat) is nailed upright to the cross-member so that the same edge of each strip is directly over the center-line of the sewer-pipe location.

The sewer grade is progressively set on these grade slats at a calculated distance above the inside bottom of the sewer pipe (the invert). The workers measure from this mark to the invert of the pipe. The distance from each mark should be some whole number of feet.

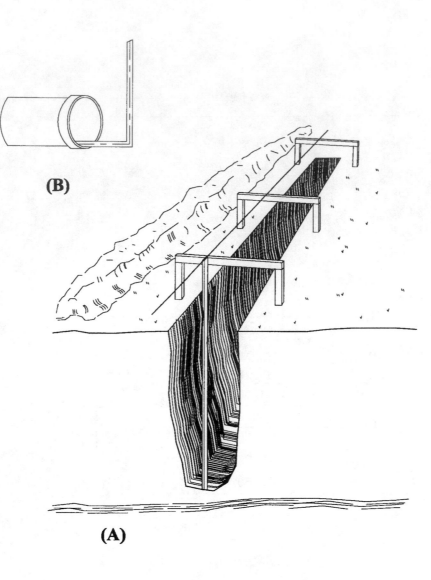

(B)

(A)

Figure 10-3 *Planning sewers*

A line, usually an iron wire, is stretched tightly along the marks to furnish a measuring base between the structures. Grade slats support the wire as shown in Figure 10-3A. The stakes and cross-members are strong and firm. The heavy black line represents a line (iron wire) on the sewer grade.

In the foreground a wooden measuring stick cut to some even foot length is shown. A short piece of wood can be fixed at 90° to the bottom of this stick to allow transfer of the measurement from the line down the stick to inside the tile for the invert elevation (Figure 10-3B).

Figure 10-4 shows a variation of this idea with heavy planks across the excavation. Plank ends are sunk firmly into the soil. Grade slats then are nailed to the planks and an iron wire is stretched across the slats for measuring down to the sewer pipe. In every case, the iron wire must be firmly anchored.

Grading

Grading an area of land usually requires either that excess soil be transferred from the site or that needed soil be transferred into the site. When the grading situation is ideal, the volume of cut actually equals the volume of fill needed for the site.

■ Determining the Amount of Earth To Be Moved

Generally, for average building, the excavation is small with a fairly uniform grade surface. If you consider the excavation to be a prism, multiplying the area of the excavation by the average height of the corners will give you a close approximate of the volume of earth you will need to move.

For a Small Area. First, you will need to find the elevation of the corners. Figure 10-5 shows an excavation site as a prism, placed on a contour map of the area. Normally, the excavation corners fall between contours. Interpolation between the contours gives you the elevation of the corners.

For instance, interpolating between the contours $A = 110.2$, $B = 116.43$, $C = 121.62$, and $D - 115.26$. Assume that the bottom of the excavation is to be at El. 106.00. If you subtract 106.00 from the above elevations, you will get the heights at the excavation corners as $A = 4.2$, $B = 10.43$, $C = 15.62$, and $D = 9.26$.

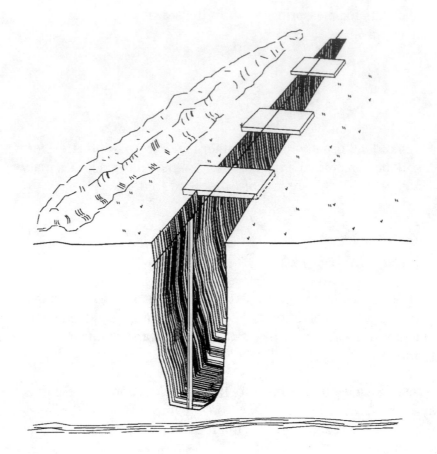

Figure 10-4 *Using stakes and cross-members*

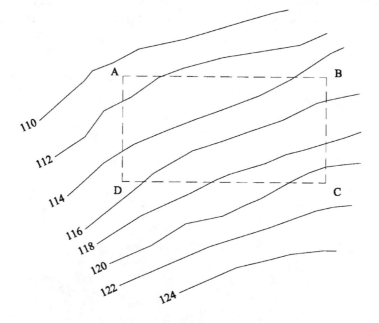

Figure 10-5 *An excavation site seen as a prism*

To interpolate these numbers, use the 20-scale of an engineer's scale. Each 1/4-inch mark represents 1 foot and each small division represents 0.2 foot of contour interval. In every case, use 1/2 inch of the scale to span between the contours. Place the rule exactly.

When you place the rule between the 110 and 1/2 contours, the first small division falls on the corner at *A*. The contour you are measuring from is 110. Then the reading is 110.2, since each small division represents 0.2 foot of contour interval.

At corner *C*, when you put the rule across the span of contours 120 and 122, the corner falls above the halfway point, which places it above contour 121, a span greater than 1/4 inch. Also, the corner is three small divisions past the 1-foot (1/4 inch) mark or, 3×0.2 inch $= 0.6$ inch. This is at the 121.6 elevation. Additionally, the corner is, if you look carefully, 0.2 distance of the space between two small divisions. This is 0.02 inch. Putting it all together, $121 + 0.6 + 0.02 = 121.62$ inches for corner *C*.

Second, take the area 20×40 feet of the excavation and multiply it by the average height of the corners to get $(4.2 + 10.43 + 15.62 + 9.26)/4 \times 20 \times 40 = 7902$ cubic feet of soil. Then, divide the number of cubic feet by 27 (the number of cubic feet in a cubic yard) to get 292.67 cubic yards of soil.

For a Large Area. For larger excavations the computation is more extensive. Figure 10-6 shows a large excavation area divided into 10-foot squares on contours at 1-foot intervals. Since the area is 60×60 feet, intervals of 1 foot give you a close control of the volumes. Suppose the bottom is at El. 110.00.

Use the method in Figure 10-5 to get the sum of the heights of the various corners. List the figures for all of the corners labeled *a:* 110.65, 116.20, 116.60, 118.64, 120.25, 122.62. Because the bottom is at 110.00, use the net height above 110.00: $0.65 + 6.20 + 6.60 + 8.64 + 10.25 + 12.62 = 44.96$.

List all corners labeled *b:* 111.50, 111.88, 1/2.47, 1/2.88, 113.49, 113.80, 114.40, 114.62. The net height is $1.50 + 1.88 + 2.47 + 2.88 + 3.49 + 3.80 + 4.40 + 4.62 = 25.04$.

And, list 115.33, 115.56, 117.52, 117.17, 117.11, 117.64, 117.83. The net height is $5.33 + 5.56 + 7.52 + 7.17 + 7.11 + 7.64 + 7.83 = 48.16$.

And, list 118.75, 119.00, 119.67, 119.69, 120.77, 121.70, 121.33. The net height is, $8.75 + 9.00 + 9.67 + 9.69 + 10.77 + 11.70 + 11.33$. The sum of all the net heights is 70.91.

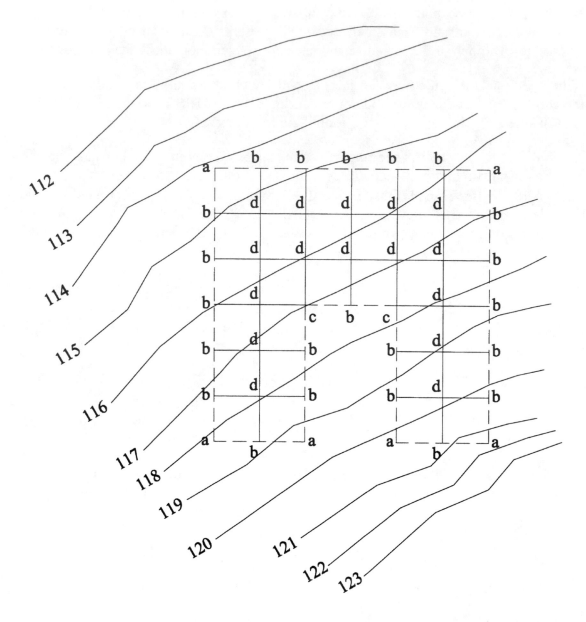

Figure 10-6 *A large excavation area*

List the corners labeled *c*: 116.00, 118.00. The net height is 6.00 + 8.00 = 14.00.

Add all the corners labeled *d*: 1/2.75, 113.80, 113.87, 114.70, 114.80, 114.84, 115.75, 115.80. The net height is 2.75 + 3.80 + 3.87 + 4.70 + 4.80 + 4.84 + 5.75 + 5.80 = 36.31.

And list 115.98, 116.80, 117.00, 117.08, 117.88, 119.00, 119.70, 120.52. The net height is 5.98 + 6.80 + 7.00 + 7.08 + 7.88 + 9.00 + 9.70 + 10.52. The sum of all the net heights is 63.96.

The total volume of the excavation is computed by the following formula: volume equals area of one square divided by 27 times the sum of the a's plus 2 times sum of the b's plus 3 times sum of the c's plus 4 times sum of the d's. Note that the Greek letter Σ is used in mathematics to read "the sum of."   Then,

volume = area of one square/27

x (Σa's + 2Σb's + 3Σc's + 4Σd's)/4.

Or, using the information from above,

volume = $100/27 \times (44.96 + 2 \times 144.11 + 3 \times 14 + 4 \times 100.27)/4$

= $100/27 \times 776.26/4 = 19,406.5/27$

= 718.80 cubic yards of soil to be excavated.

For a Very Large Area. The preceding methods of calculating volumes are called *unit area* or *borrow-pit* methods. But for even larger areas, they are too time consuming.

To measure areas of plane figures for extremely large areas, use an instrument called a *planimeter*. You can find one at any surveyor's supply (instructions for its use come with it).

∎ Setting Cut-and-Fill Grade Stakes

In order to grade, you must set cut-and-fill grade stakes. Set them at a full number of feet above or below the finished elevation you want. Mark the cut or fill plainly on each stake: "Cut 2 feet", or "Fill 2 feet."

If you want the finished grade to be level, set up and find an HI on a control point at the finished grade. Depending on the natural grade, you may be able to set all the stakes from the one HI. If you need more than one setup, you will need a backsight and a new HI.

When you want to slope the grade in a certain direction, set the lines of stakes parallel to the direction of slope. Set one row of stakes in the direction for the slope, then set cross-rows at right angles. All the stakes in any one row are at the same elevation.

When you want the finished grade to slope in two directions, use the same method, especially if the directions are at right angles to each other.

Your work is more complicated if the two directions of slope are not at right angles. Use several controlling points and set stakes in a straight line between those points.

Necessary Curves

You do not need to grade all hills. Due to the natural landscape of hills and valleys, some curves are actually necessary for constructing streets and roads. There are various classes of horizontal curves used for roads, but the *simple*, or *circular curve*, is used for most street construction.

■ Simple Curves

The straight portions of a street at each end of a curve are called *tangents*. Each is tangent to the curve that connects them. The curve introduces an abrupt change from the tangent direction. Where speeds are high, a *transition (spiral)* curve connects the tangent to the circular curve. On city streets the speed limit is such that the change can be made from tangent to circular curve.

Figure 10-7 shows a simple curve connecting two tangents. The curve has a uniform radius for its entire length, and the center is at 0.

AB and *CD* are the tangents (straight roadway) connected by the curve (arc) *BC*. The radius is *OB* or OC.

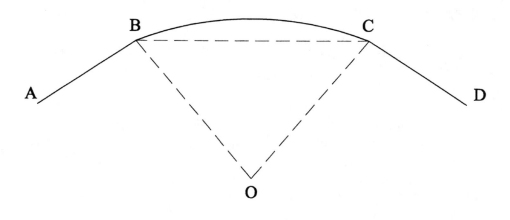

Figure 10-7 A simple curve connecting two tangents

The degree of sharpness, or form, of a simple curve is designated by the radius. The radius usually is too large for practical layout by swinging a tape from the center. But the value of the radius is used to compute other values necessary to lay out the curve.

The sharpness of a curve is given by the term *degree of curve*. This is the angle formed at the center of the curve by two radii passing through points on the curve 100 feet apart. Some engineers and surveyors assume the 100-foot distance lies along the arc of the curve. Others assume the 100-foot distance lies along the chord that connects the two points.

In Figure 10-7 this distance is, in the first case, along the arc of the curve from *B* to *C*. In the second case, the distance is along the chord of the arc, the broken line from *B* to *C*.

For the field work, you will already know the route along which the 100-foot distance is to be measured. This is decided during the planning stage.

In Figure 10-8 the tangents *AB* and *MN* are connected by a simple curve *EF*. The points *G*, *J*, and *L* on the curve are 100 feet apart. In this case, the 100-foot distance is along the chord (broken line) from *G* to *J* and from *J* to *L*.

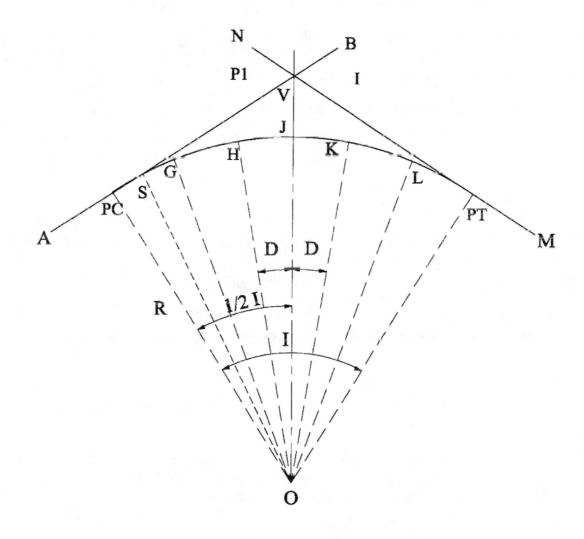

Figure 10-8 *The measure (in degrees) of angle D, equals the measure of the curve*

The degree of curve is the angle *D* measured at the center *O* of the curve between the radii *OG* and *OJ*, and between *OJ* and *OL*. The measure, in degrees, of angle *D* is the measure, in degrees, of the curve.

■ Radius and Curve

Use computations based on a 100-foot chord to determine the relationship between the radius and the degree of curve. For instance, a line, *OH*, drawn from the center *O* of the curve to a point bisecting the 100-foot chord *GJ* is perpendicular to the chord. The angles *GOH* and *HOJ* each equal one-half *GOJ*.

The triangle *GOH* is a right triangle subject to the trigonometric laws for right triangles. In right triangle *GOH*, sin *GOH* = *GH/OG*.

Line *OH* divides the angle *GOJ* and the chord *GJ* equally, and *GH* = 50 feet. Line *OG* equals the radius shown as *R*. Since *GOH* = D/2, sin 1/2D = 50/R, as given differently in the preceding paragraph (D = degree of curve and R = radius of the curve, in feet).

The formula can be changed to find the radius when the degree of curve is known. Interchange sin 1/2*D* and *R* to have *R* = 50/sin 1/2*D*.

■ Chords and Angles

There can be any number of 100-foot chords and angles (D) in a curve, depending on the curve length. The point E is the beginning of the curve and is called *point of curve* (*PC*). The point F is the end of the curve and is called *point of tangent* (*PT*).

The intersection of the tangents *AB* and *MN* is called the point of *intersection* (PI), and sometimes the *vertex*, denoted by the letter V. The external angle, I, formed by the tangents is called the *intersection* angle.

The tangent *AB* and radius *R (OE)* are perpendicular, as are *NM* and *OF*. Therefore, the intersection angles *NVA* and *BVM* are both equal to the total central angle *EOF* between the radii *OE* and *OF* passing through the endpoints of the curve.

Since, by geometry, the lengths of two tangents to a circle from a common point outside the circle are equal, *EV* = *FV*. Then, *PC* and *PT* are equidistant from *PI*. Either distance, *PC* to *PI* or *PT* to *PI*, is called the *tangent distance*, usually designated by letter *T*.

■ Tangent Distance

The tangent distance, the radius, and the central angle of the curve are directly related as you can see in Figure 10-8. The line *OV* bisects the angle *EOF* and forms the right triangle *OEV*. Then tan *EOV* = tan 1/2*EOF* = *EV/EO*, and *EV* = *EO* tan 1/2*EOF*.

EV = *T*, the tangent distance. *EO* is the radius and *EOF* is the total central angle (I). Substituting in the above formula, T = R tan 1/2*I*. Note that tan 1/2*EOF* (also tan 1/2*I*) = tan *EOV*, since *EOV* is 1/2*EOF*. This formula is true for all curves with a degree of curve based on either a 100-foot chord or a 100-foot arc. Since R is slightly different for a chord and an arc (*OH* is slightly shorter than *OE*), the correct value of *R* must be used.

For example, suppose two tangents intersecting at an angle of 26°40' are to be joined by a 2°30' curve. Compute the tangent distance for a curve for which the degree is measured by a 100-foot chord. Since *T* = R tan 1/2*I*, you must first find *R*. Use the formula you learned for *R* = 50/sin 1/2*D*.

$R = 50/\sin 1/2D - 50/\sin 1/2\ 2°30'$

$= 50/\sin 1°15' = 50/.02181$

$= 2292$ feet.

Then,

$T = 2292 \tan 1/2I = 2292 \tan 1/2\ 26°40'$

$= 2292 \tan 13020' = 2292 \times .23700$

$= 543.2$ feet, the tangent length.

If the degree of curve is given as the angle between two radii intersecting a curve at the ends of a 100-foot arc, the degree of curve and its radius have a relationship determined by the facts of a circle. Remember: This curve is based on measurement along the arc, not the chord. The total central angle for a circle is 360° and it is subtended by the complete circumference of the circle. The degree of curve is an angle subtended by an arc or 100 feet.

Based on these facts, an exact proportion can be written for R. That proportion is:

$D/360 = 100/2\pi R.$

Then,

$2\pi D/360 = 100/R,$ and,

$R = (100 \times 360)/2\pi D = 36,000/6.2832D = 5729.56/D \quad R = 5729.56/D$

$\pi = 3.1416$, and

$2\pi = 6.2832$

If the radius is given, transpose to find D.

$D = 5729.56/R$

Again, the length of a curve is based on the degree of curve for either a 100-foot chord or a 100-foot arc.

When based on a 100-foot chord, the length of the curve is the sum of all the chords from the PC to the PT. In Figure 10-8 this sum along the chords is $EG + GJ + JL + LF$. Only two 100-foot chords are shown in this figure, GJ and JL, and two short chords, EG and LF.

The drawing is for illustration only: The number of 100-foot chords in a curve actually depends on its length.

For the short chords it is generally assumed that the difference between the station numbers at the ends of the subchord is in the same ratio to 100 feet as the central angle subtended by the subchord is to the degree of curve (D). If c' represents the difference between station numbers, then

$c'/100 = EOG/D$, and $EOG = (c'/100) \times D$

There are several techniques for finding these answers, but equations based on plane geometry and trigonometry give answers for any practical purpose. For instance, the perpendicular bisector OS of the subchord forms the right angle ESO. Then, sin $ESO = ES/R$ and, since $ES = 1/2c'$, sin $EOS = 1/2c'/R$, and $1/2c' = \sin EOS \times R$. c' equals two times the answer from the equation.

■ Length of Curve

The length of a simple curve with the degree of curve based on either a chord or an arc of 100 feet can be found by the formula $L = 100/D$. The length of a curve with $I = 26°40'$ and $D = 2°30'$ is $L = 26.666/ 2.5 \times 100 = 1066.64$ feet. Decimal degrees for $26°40' = 26.666$. Forty minutes is two-thirds of sixty minutes: $2°30' = 2.5°$.

■ Laying Out Curves

Due to topographic conditions, you must usually lay out a curve without locating the *PI* on the ground. Before the field layout begins, establish the *PI* on a map of the area and a curve selected to fit the conditions. This curve will have a tangent distance. The station of the *PC* is found by subtracting the tangent distance from the station of the *PI*. The station of the *PT* is found by adding the length of the curve to the *PC*.

Consider the example with the tangent distance 543.2 feet and the length of curve 1066.64 feet. Subtract the tangent distance 543.2 from whatever station number is assigned on the area map for the *PI* and find the *PC*. Add 1066.64 to the *PC* station number to find the *PT* station number.

■ Setting Stakes

To stake a curve on location, a stake is set at each full station of 100 feet. Stakes can be set at intermediate points when necessary. Topographic conditions can make it impossible to set a stake at 100-foot stations. The stakes are then set at intervals of 50 feet, or even 25 feet. A common rule places stakes at 100-foot intervals for a degree of curve less than 8°, 50-foot intervals for a degree of curve from 8° to 160 and 25-foot intervals for curves tighter than 16°

If the measurement for degree is based on a 100-foot arc, the distance from the *PC* to the *PT* is along the arc. Then, when staking out the curve, if distances are measured along chords joining points on the curve, you must determine a chord for 100 feet of arc. Also, you must find the chord length for any arc that falls short of 100 feet.

The formula for finding the chord of a 100-foot arc is: Length of chord for a 100-foot arc, in feet, is $c = 2R \times \sin 1/2D$ (c = length of chord, R = radius of curve, in feet, and D = degree of curve).

For an arc less than 100 feet long, $c = 2R \times \sin aD/200$ (c = length of subchord for any given arc, R = radius of the curve in feet, a = length of the arc, in feet, and D = degree of curve). Since, for the average street, curves based on the 100-foot chord are adequate, the derivation of these formulas is not given.

Curves with degrees based on 100-foot chords are somewhat flat. For such curves, it is considered sufficient to set stakes at full stations only. The subchords then fall at the *PC* and *PT*.

The length of the subchord beginning at the *PC* and ending at the first full station on the curve is taken as the length obtained by subtracting the plus footage in the *PC* station number from 100. For instance, if the station number for the *PC* is given as 4 + 71, the sub-chord is 500 − 471 = 29 feet.

The length of the subchord from the last full station on the curve to the *PT* is the same as the plus footage in the *PT* station number. For instance, assume in Figure 10-8 that the *PT* station number is 7 + 56. Then the station number of point L is 7 + 00 and the subchord length is 56 feet.

Where 100-foot chords are used and stakes are set between the regular 100-foot stations on a sharp curve (exceeding 8°), you must make an allowance for the difference between the length of each sub-chord and the station interval between the ends of the sub-chord. For 50-foot intervals, each 50-foot station is set at the center of the arc between adjacent full stations (100-foot interval). For 25-foot intervals, each 25-foot station is set at the center of an arc measuring one fourth of the arc between adjacent full stations.

In Figure 10-9, chord *AE* is 100 feet long with stakes to be set on the proposed curve at 25-foot intervals. Since the length along the curve is greater than 100 feet, the sum of the four chords *AB*, *BC*, *CD*, and *DE*, is greater than 100 feet. You must set the stakes at points dividing the arc into four equal subarcs.

■ Subchords

Since each subchord is longer than 25 feet, it is necessary to find the subchord length. Set the difference, c_s, between the station numbers at the ends of the subchords in the same ratio to 100 feet as the central angle subtended by the subchord is to the degree of curve (*D*). Then $c_s/100 = AOB/D$ where c_s = difference in feet between the station numbers at the ends of the subchord.

The central angle *AOB* subtended by the subchord is $AOB = c/D$. If a perpendicular bisector *OM* is established from *O* to the subchord *AB* and the right triangle *AOM* is solved the same as for a full chord, then, $c = 2R \sin c_s D/200$ (where c = actual length in feet of the sub-chord, R = radius in feet of the curve, and D = degree of curve).

For example, find the actual length of the sub-chord between station 10 + 00 and station 10 + 25 on a 14° curve, the degree being based on a 100-foot chord.

First, find the radius R. The perpendicular bisector *OC* of the chord and the arc forms one side, *OF*, of the right triangle *OFA*. $AF = 1/2AE = 50'$. Also, angle $AOF = 1/2D$.

Then, in the right triangle *OFA*,

R = 50/sin 1/2D = 50/sin 7° = 50/.12187 = 410.273.

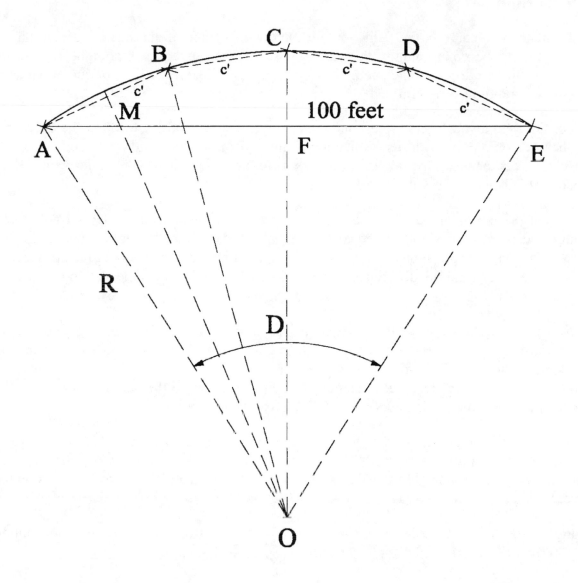

100 feet

Figure 10-9 Stakes will be set at 25-foot intervals

And,

$$c = 2R \sin c_s D/200;$$

$$c = 2 \times 410.273 \times \sin (25 \times 14)/200$$

$$= 820.546 \times \sin 1.750 = 820.546 \times 0.030538$$

$$= 25.05 \text{ feet,}$$

the actual length of the subchord *AB*, *BC*, and so on.

■ In the Field

To stake out a curve in the field, you must know the degree of curve and the positions of the tangents to be connected by the curve. Usually the tangents, as straight streets, are located to suit both topographic and building conditions. Then the degree of curve that suits both these conditions is determined.

For the field layout, the simplest and most convenient method for locating the endpoints of subchords on a curve is to use deflection angles. Using the tangent as a baseline and the transit set over the *PC*, turn deflection angles to locate the point where the forward end of each chord falls on the curve.

By the principles of geometry, an angle between a tangent to a circle and any chord through the point of tangency is equal to one half of the central angle subtended by that chord. The central angle subtended by any chord on a simple curve is equal to the product of the degree of curve and the difference, in stations of 100 feet, between the station numbers at the chords ends.

If *l* equals the distance in stations between the chord ends and *D* is the degree of curve, the central angle subtended by the chord equals $2D$. According to the geometric principle given above, then $d = lD/2$ where d = deflection angle from the tangent at the *PC* to any point on the curve, 1 = the difference, in stations of 100 feet, between the station numbers of the *PC* and the given point on the curve, and *D* = degree of curve.

The deflection angle is the same for a degree of curve based on either a 100-foot chord or a 100-foot arc for a given length (*l*). For example, the *PC* of a 4° curve is at station 9 + 32. Compute the deflection angle from the tangent at the *PC* to each of these given points on the curve: Stations 10 + 00, 11 + 00, 12 + 00, 13 + 00, and 13 + 77.

Figure 10-10 shows the layout. The distance in stations from the *PC* (point *B*) to station 10 + 00 (point C) is 10.00–9.32 = 0.68 station. The deflection angle is:

$$d = lD/2 = (0.68 \times 4)/2$$

$$= 1.36°, \text{ or } 1°21'36".$$

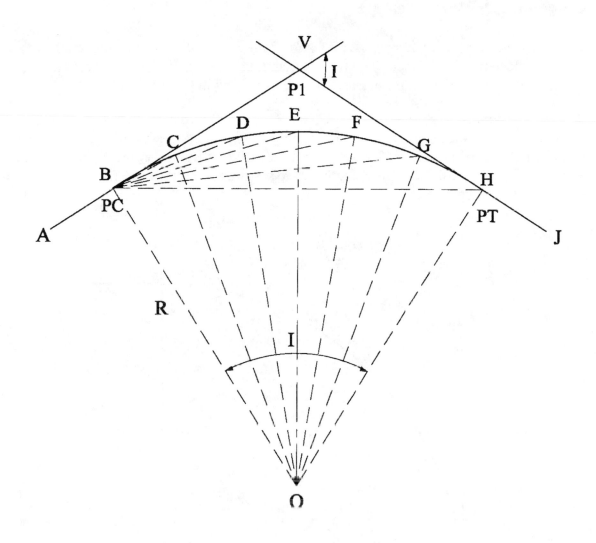

Figure 10-10 *Deflection angles*

The distance in stations from the *PC* to station 11 + 00 (point *D*) is 11.00 − 9.32 = 1.68 stations. d = (1.68 × 4)/2 = 3.36°, or 3°21'36". For station 12 + 00 (point E), I = 2.68 stations and d = 5.36°, or 5°21'36". For station 13 + 00 (point *F*), I = 3.68 stations and d = 7.36°, or 7°21'36". For station 13 + 77 (point *G*), l = 13.77 −9.32 = 4.38 stations. d = (4.38 × 4)/2 = 8.76°, or 8°45'36".

■ Street and Alley Intersections

Generally, modern standards set the radius for a street and alley intersection at 20 feet and for street intersections at 30 feet. To lay out an arc at a street and alley intersection, extend each curb line through a point of intersection such as point *F* in Figure 10-11.

Drive stakes on the curb lines at *A*, *B*, *C*, and *D*. Mason cord stretched between stakes *A* and *C* and *B* and *D* intersects at point *F*. Set points *X* and *Y* 20 feet from point *F*. Stretch two tapes one each from points *X* and *Y*. Where the 20-foot marks on the tapes coincide is the point 0. Then, *R* = 20 feet. Describe the 20-foot radius from point *O*. Determine the 30-foot street radius the same way.

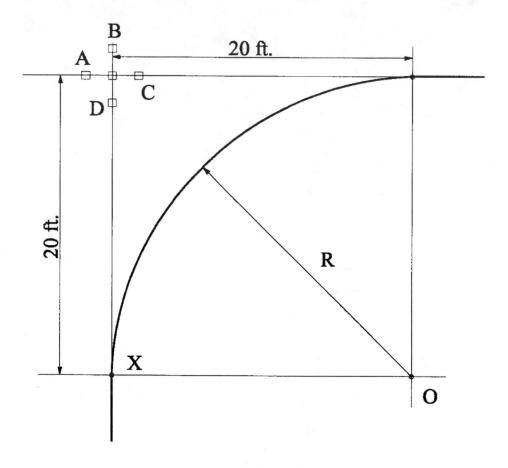

Figure 10-11 Street and alley intersection

If a transit's set at point *Y*, you can extend the curb line by plunging the telescope. You can make a 20-foot measurement from *Y* to *F* and set a tack in a stake at point *F*. Then you can turn a 90° angle left from point *F* and repeat the procedure to set point *O*. The arc is described as previously.

■ Other Curves

A *vertical* curve (sometimes called a *profile* curve) is used to control change in grade in hilly terrain. When an upward slope changes to a downward slope (peak curve), or when a downward slope changes to an upward slope (sag curve), a vertical curve is used.

Parabolic curves are used to construct vertical curves in streets to provide easy transition in a vertical direction. The parabola is a plane curve formed such that every point on it is equally distant from a fixed point, called a focus, and straight line, called the *directrix*.

A parabolic curve such as that in Figure 10-12, is formed when a point traces a curve, moving in a path such that the point always remains an equal distance from a focal point and a straight line. The points *a'*, *b'*, and *c'* lie on the curve. The distance *aa'* is equal to the distance from *a'* to the focus. The same is true for the distances *bb'* and *cc'*. The point moves from the vertex to *a'* to *b'* to *c'* (and onward) to trace the curve. The vertex also is an equal distance from both the focus and the *directrix*.

■ The Tangent-Offset Method

A vertical curve is tangent to both grade lines and usually is 50 to 200 feet long. The layout is made by offsets from the tangent. The offset distance from the tangent varies as the square of the distance along the tangent. This is called the *tangent-offset method* and requires only simple arithmetic to construct.

The tangent-offset method is applicable to curves that have either equal tangents or unequal tangents. Since the slopes connected by the curve may have different grades, or topographic conditions may make it necessary to run a greater distance forward from the curve vertex than from the curve vertex backward, a curve with unequal tangents must be fitted to the condition.

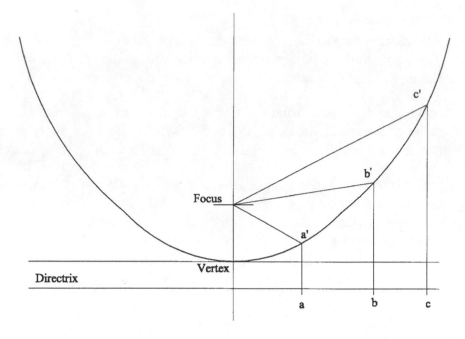

Figure 10-12 *A parabolic curve*

Equal-Tangent Curve. First, consider an equal-tangent curve. A downward slope is marked minus and an upward slope is marked plus. In Figure 10-13 the downward slope is minus 0.03 (−0.03) and the upward slope is plus 0.04 (+0.04). The two slopes intersect at point V (the PI) at an elevation of 695.46. You find this elevation and the slope grades by direct leveling.

A 400-foot long vertical sag curve (parabolic) in horizontal projection extends from A to B. Since all measurements of length in this plane are horizontal projections, lengths ADB, ACB, and AVB are all equal to horizontal projection A_1B.

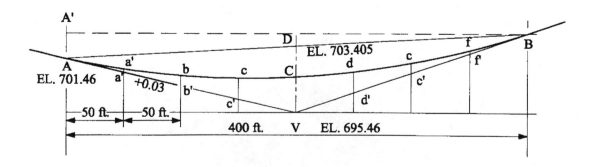

Figure 10-13 *An equal-tangent curve*

Point C is midway between points V and D. The point V is midway between A and B. AV and BV are each equal to 200 feet. The elevations along the lines of intersection (existing grade) are as follows.

$A = 695.46 + (200 \times 0.03) = 695.46 + 6 = 701.46$

$a' = 695.46 + (150 \times 0.03) = 695.46 + 4.5 = 699.96$

$b' = 695.46 + (100 \times 0.03) = 695.46 + 3.0 = 698.46$

$c' = 695.46 + (50 \times 0.03) = 695.46 + 1.5 = 696.96$

$V = 695.46$ (established by direct leveling)

$d' = 695.46 + (50 \times 0.04) = 695.46 + 2.0 = 697.46$

$e' = 695.46 + (100 \times 0.04) = 695.46 + 4.0 = 699.46$

$f' = 695.46 + (150 \times 0.04) = 695.46 + 6.0 = 701.46$

$B = 695.46 + (200 \times 0.04) = 695.46 + 8.0 = 703.46$

Parabolic Curve. The following computations establish the parabolic curve.

$D = (701.46 + 703.46)/2 = 1404.92/2$

$= 702.46$ (point D is midway between A and B and the elevations are averaged).

$C = (703.46 + 695.46)/2 = 699.46$ (averaged).

$VC = 699.46 - 695.46 = 4.0$ feet.

Lines AV and BV are tangent to the parabolic curve at points A and B. The y (ordinate) distances aa', bb', and so on, are proportional to the squares of their distances from the point of tangency A. Point a' is $50/200 = 1/4$ the distance from point A to point V. Point bb' is $100/200 = 1/2$ the distance from point A to point V. VC, the midpoint of the curve, is 4.0 feet.

To be a true parabola, aa' must equal ff', bb' must equal ee', and so on. Then,

$aa' = ff' = (1/4)^2 \times 4 = 1/16 \times 4 = 0.25$ feet.

$bb' = ee' = (1/2)^2 \times 4 = 1/4 \times 4 = 1.0$ feet.

$cc' = dd' = (3/4)^2 \times 4 \ 9/16 \times 4 = 2.25$ feet.

Adding these distances to the elevations on the grades gives you the elevations at corresponding points on the curve.

$a = $ El. $a' + 0.25$ feet $= 699.96 + 0.25$ feet $= 700.21$ feet

$b = $ El. $b + 1.0$ feet $= 698.46 + 1.0$ feet $= 699.46$ feet

$c = 696.96 + 2.25$ feet $= 699.21$ feet

$d = 697.46 + 2.25$ feet $= 699.71$ feet $(cc' = dd')$

$e = 699.46 + 1.0$ feet $= 700.46$ feet

$f = 701.46 + 0.25$ feet $= 701.71$ feet

If Figure 10-13 were turned so that V is on top (peak curve) and the grades (tangents) run downward, the curve would be the same. The offsets, however, are subtracted from the elevations of points on the grade lines (tangents).

Unequal-Tangent Curve. To compute for an unequal-tangent curve, establish a curve over each tangent. In Figure 10-13, a 400-foot curve was computed and explained. In Figure 10-14 the 400-foot curve from station 21 to station 25 is a down-slope curve. The up-slope curve from station 25 to station 31 is 600 feet. This is an unequal-tangent curve. The length of curve is chosen to fit topographic conditions.

Stations 21 through 31 lie on the level line AB. The center of the down-slope curve is at station 23. The center of the up-slope curve is at station 28. Where these stations fall on the tangents, connect them with the line CD.

Compute the elevation of point VC from the known elevations of stations 23 and 28. Thus, $703.84 - 703.24 = 0.60$, the difference in elevation between the two stations.

Proportionally, there are five station intervals, 500 feet, between these two stations. The distance from C to VC is two stations of the five (2/5 of 5), and the distance from VC to D is three stations of the five (3/5 of 5).

Then, 2/5 times the difference in elevation (0.60) plus the elevation at C yields the elevation at VC ($703.84 - 703.24 = 0.60$). The calculation is:

$2/5 \times 0.60 + 703.24 = 1.2/5 + 703.24 = 703.48$

Station 23 is the low station and the 2/5 difference must be added.

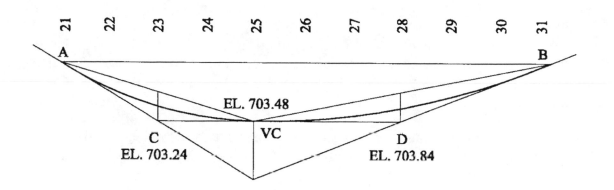

Figure 10-14 An unequal-tangent curve

If station 28 elevation is used, the 3/5 difference must be subtracted. Here, $703.84 - 3/5 \times 0.60 = 703.84 - 1.8/5 = 703.48$.

A 400-foot curve exists from A to VC and a 600-foot curve exists from B to VC. Both curves are tangent to CD at point VC and, therefore tangent to each other which yields a smooth transition from the down-slope to the up-slope curve.

Marking The Low Point. Conditions such as motorist sight distance, drainage, clearance beneath overhead structures, and other situations may require the location and elevation of the low point on a vertical curve (or high point).

In Figure 10-13 point C would be the low point if the slopes were the same. Since the slopes are different, the low point lies left of C along the line of lowest grade (0.3 as compared to 0.4).

The horizontal distance from the beginning point of a curve to the low (or high) point of the curve is $Lg/g_1 - g_2$. L is the horizontal projection of the total length of the curve, or 400 feet. The grade percentages at the beginning and end of the curve are g_1 and g_2, respectively, or -0.03 and $+0.04$. Review the plus and minus grades from the equal-tangent curve discussion.

In Figure 10-13 the low point L is left of C. The horizontal distance of $AL = (400 \times -0.03)/[(-0.03) - (+0.04)] = (400 \times -0.03)/0.07 = -12/-0.07 = 171.43$ feet. (Note that $400 \times (-0.03) = -12$, and $(-0.03) - (+0.04) = -0.07$. If needed, review the operation of sines in the appendixes section.)

Find the elevation of point L' on the tangent line to find the elevation of the point L on the curve. The elevation of A is 701.46. Then, $701.46 - (0.03 \times 171.43) = 696.32$, the elevation of point L'.

From the calculation of the 400-foot curve, the distance $VC = 4.0$ feet. The horizontal distance $AV = 200$ feet. Since a vertical offset is proportional to the square of its distance from the point of tangency, $L'L = (171.43/200) \, 2 \times 4.0 = (0.85715)^2 \times 4.0 = 0.734706 \times 4.0 = 2.9388$, for example, 2.94.

Since the elevation of point L' plus 2.94 equals the elevation of point L, $696.32 + 2.94 = 699.26$, the elevation of point L. This point is $699.46 - 699.26 = 0.20$ inches lower than point C.

To stake a vertical curve, set the stakes at the points of calculated elevation. Place a mark on the stakes at the exact elevation and mark the elevation above the mark. Mark the amount of cut (or fill if it is a crest curve) on the stake for the grading contractor.

CHAPTER 11

Surveying Notes

Original surveying notes are those you record while working in the field. These notes are your permanent record of the field work. As previously noted, you should never erase anything on a sheet of surveying information. Every surveyor eventually gets called to court to be a witness in a land dispute. And in court only original survey notes are acceptable.

Also, the legal description for a deed is made from your survey notes. The deed is recorded in the county recorder's office and information therein, especially the legal description, may be referred to in the future when land transactions are made. So, your notes must be correct.

You must record every measured distance, angle, or elevation that you take. If you discover an incorrect measurement, just draw a thin line through it. That way, it is invalidated but still legible. Never erase. To invalidate a paragraph or entire page, draw diagonal lines across it from opposite corners. Also, write the word VOID in a clear space beside the paragraph or on the page.

When you are surveying, the most important factor is accuracy in measurement. Your surveying accuracy is then reflected in accurate field notes. Since sketches and tabulations are a part of the field notes, they too must clearly and accurately present your work.

In Chapter 12, "Special Problems," the problem of missing measurements is discussed. This is a problem that reaches to the very heart of surveying note integrity. When you complete any series of measurements, never assume your notes are complete. Check them carefully for missing measurements and relative information before you leave the site.

Figure 9-5 shows sample facing pages from a surveyor's field book. Such pages provide an excellent format for taking notes for an average survey. The structured format enhances your accuracy and helps to keep everything legible. It also helps to use a hard lead pencil, 3-H hardness or harder, to make easy reading. Occasionally, your surveying field books will have to withstand damp conditions; soft lead will smudge but hard lead stays clear and decipherable.

Every surveyor develops individual habits for note taking and arranging. Generally, you will start by making a sketch of the area to be surveyed (though in some field work it will make sense to make the sketch as the work progresses). Make the sketch with enough space for each measurement or station you are planning. Place a north sign on each sketch. Also, draw the sketch large enough to write clearly.

As you take each measurement, tabulate it. Read the tape, transit scale, or level carefully. Never crowd the notes. As you have seen in this book, list significant zeros (an example is using 10.30 instead of 10.3). Place a zero before the decimal point for numbers less than one (though some surveyors do not place a zero before decimal trigonometric functions).

If situations develop that you cannot explain by sketches or notes, give a longer description. Descriptions are often given for monuments, trees, boulders, fence corners, hills, or highway structures involved in a survey. You can put descriptions on the pages immediately following the ongoing work and make references to related sketches or tabulations.

Certain items are a part of every book of surveys and of every survey. First, the field books should be marked with the year and month of beginning entries and the month and year of completion entries. Keep your field books in a safe place.

Second, for each survey list the date and time the survey begins and the date and time it ends. Leave out the lunch period by making four recordings of time. The amount of time required to do a survey is a useful piece of information.

Third, always list the weather conditions. If conditions change during the survey, list the time and type of change. Whatever the weather is, it will somehow affect the accuracy of the survey, especially since surveys are performed in extremes of temperature, with high wind velocity, and during rainstorms.

Fourth, list the name and duty of each party member.

Fifth, list the type of instrument you are using and, if applicable, the number. Note whether the tapes you are using are steel or plastic. Or, if distances are electronically measured, note that fact. When a level rod is used list the type and, if applicable, the number.

Sixth, if you use a company vehicle, list its make, type, and model.

It is best to begin each day's work on a new page. You could continue on the same page if the survey requires a lot of detail, but be sure to indicate where data for the new day begins.

CHAPTER 12

Special Problems

A particular problem of surveying occurs when measurements are missing from survey notes or a deed. Usually, the term *missing measurements* refers to missing lengths and/or bearings of polygons mapped in a survey.

Double-check all of the work you have done before you pack up and leave the site. Still, even the best of surveyors will occasionally be faced with one of the following problems.

1. The length of a side is missing. The bearing of a side is missing. Both the length and bearing of one side are missing.

2. The length of one side and the bearing of another side are missing.

3. The lengths of two sides are missing.

4. The bearings of two sides are missing.

To solve the problem of the missing measurement, you need to understand and be able to use surveying mathematics to find it, given the measurements you do have at hand. This chapter describes the steps you take to find the needed measurements.

Finding Missing Measurements

■ Solving for Condition 1

To solve for any of the missing measurements listed under condition 1, use the method discussed in relation to Figures 4-5 and 4-6. Simply tabulate and compute the latitudes and departures of all the values you have.

■ Solving for Condition 2

Figure 12-1 represents the problem in condition 2. A length only is given for course *BC* and a bearing only for course *CD*.

Take these two courses from the survey by establishing a new course, measurements unknown, from *B* to *D* (*BD*). This gives you a four-course calculation with the error of closure largely apparent since a length and a bearing are missing from one side of the four-sided figure. The difference in the sums of the latitudes and departures is the latitude and departure of the unlabeled side.

The square root of the sum of the latitude squared and the departure squared (course *BD*) is the length (see Figure 4-6, course *BC*).

The departure divided by the latitude gives you the tangent of the bearing. The arc tan of this tangent value gives you the bearing. The direction of the bearing is given by the short-sum column. For course *BD* it is north and east (Figure 12-2).

Set up the four courses of this survey as shown. Course *BD* has the blank columns you need to fill. Compute the latitudes and departures for courses *AB*, *DE*, and *EA*, then total the columns.

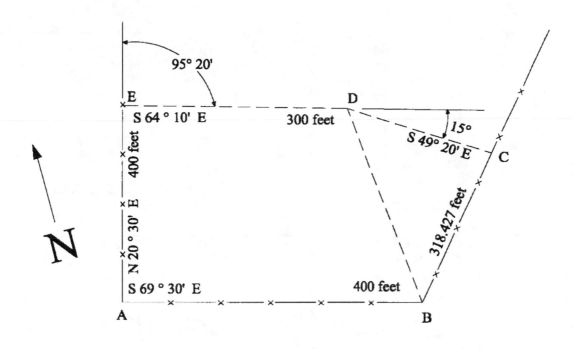

Figure 12-1 Condition 2

Note the top figures in the last horizontal row: latitudes 130.726 and 514.751, departures 374.669 and 410.839. Balance the latitudes: 514.751 − 130.726 = 384.025. This is the latitude for course *BD*. Write it in under North. Balance the departures: 410.839 − 374.669 = 36.170. This is the departure for course *BD*. Write it in under East. Finally, find the missing measurements you need by computing the bearing and length for course *BD* as shown in Figure 12-2.

Another condition is shown in Figure 12-3 where the bearing of line *KL* and the length of line *LM* are missing from the measurements of a survey.

Establish line *KM* to exclude lines *KL* and *LM*, and find the measurements of line *KM* by the method given above for Figure 12-1. The figure is similar to Figure 4-5.

Since line *KM* now has bearing and length, you can draw the triangle of Figure 12-3A to find bearing *KL* and length *LM*. This is an oblique triangle problem. What can you add to what you already know to determine the method of solution? Since two sides are known, if you can find an angle opposite one of them you can apply the law of sines.

Course		Distance		Latitudes		Departures	
No.	Bearing	Cos	Sin	North	South	East	West
AB	S 69°30' E	400 .35021	00 .93667		140.082	374.669	
BD	N 5°22'50" E	385 Tan Bearing = .094187	725	384.025		36.170	
DE	N 64°10' W	300 .43575	00 .90007	130.726			270.019
EA	S 20°30' W	400 .93667	00 .35021				140.082
				130.726 384.025 514.751	514.751 130.726 384.025	374.669 36.170 410.839	410.839 374.669 36.170

Figure 12-2 Notes on condition 2

The north sign at *K* shows course *KM* as N5'20'E. Place a north sign at *M* and the course *KM* becomes S5'20'W. Look at this in the conventional quadrant form in Figure 12-3B.

The bearings are given as back bearings with each course placed in its proper quadrant. If you add the two bearings you get the angle *LMK*, (50°00') + (5°20') = 55°20', view Figure 12-3A. Then the triangle, Figure 12-3C, has two known sides and a known angle opposite one of them. You can solve this triangle by the law of sines:

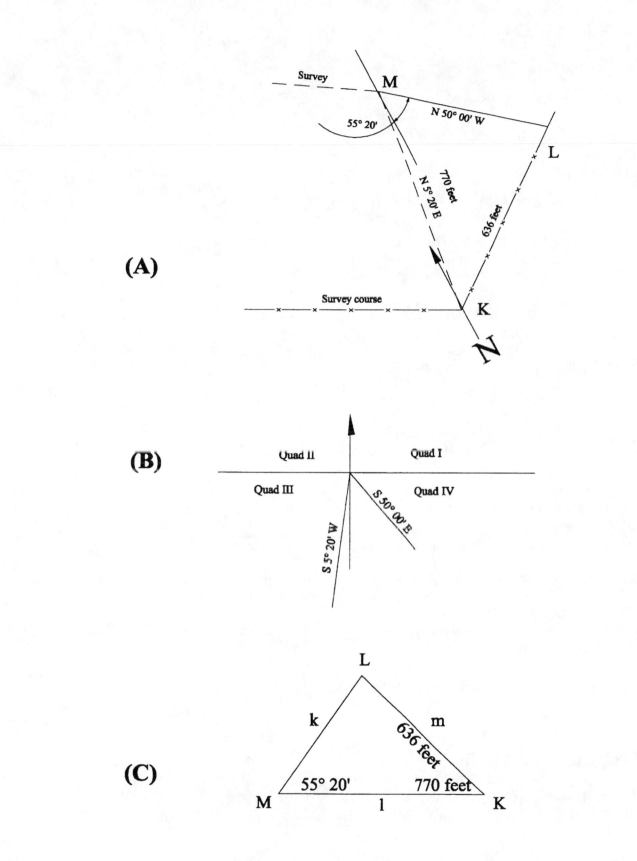

Figure 12-3 *Finding the measurements of line KM*

$m/\sin M = l/\sin L$;

636/sin 55°20' = 770/sin L;

sin L = 770 sin 55°20'/636 = .99576.

Arcsin .99576 = 84°43'41", angle L.

Therefore,

Angle $K = 180° - (L + M)$

= 180°− [(84°43'41") − (55°20'00")]

= (179°59'60") − (140°03'41") = 39°56'19".

By the north sign at K, Figure 12-3A, you can tell that bearing KL is N39'56' 19"E. Add this to the bearing of KM,

(39°56'19") + (5°20'00") = 45°16'19".

The bearing of course KL is N45'16'19"E. By the law of sines:

$m/\sin M = k/\sin K$;

636/sin 55°20' = k/sin 39°56'19"

Next, convert to decimal degrees:

636/55.33 = k/39.9386

Then, find sine values:

636/.822442 = k/.641966

And finally, solve for k:

k = .641966 × 636/.822442 = 496.4 feet

■ Solving for Condition 3

Figure 12-4 represents condition 3. In this survey the lengths of two sides, BC and CD are missing. Figure 12-5 gives the calculations made for latitudes and departures.

To find the missing lengths start by drawing a line from *B* to *D* as shown in Figure 12-6 to eliminate the courses *BC* and *CD*. Do the calculations shown in Figure 12-7 to find line *BD*.

When you draw line *BD*, you will create a triangle with lines *BC* and *CD* as shown in Figure 12-8A. It shows all of the values you already know: the bearings of the three sides and the length of *BD* as you computed it in Figure 12-7. You still need to find the lengths of *BC* and *CD*.

When you draw a figure such as this, include north signs as shown at points *B*, *C*, and *D*. This way each quadrant in which a line lies is given with respect to a particular north sign. By doing this, you can easily keep lines and bearings in context.

At *B* both bearings (*BC* and *BD*) are in quadrant IV running south and east from north. The angle at *B* is bearing *BD* minus bearing *BC*: 81° − 67.77° = 13.230. Mark this in as shown in the view Figure 12- 8B.

At *C* use the back bearing of *BC* and you will get the angle between north and *BC* as 67.77°. The angle between north and *CD* is 78.90°.

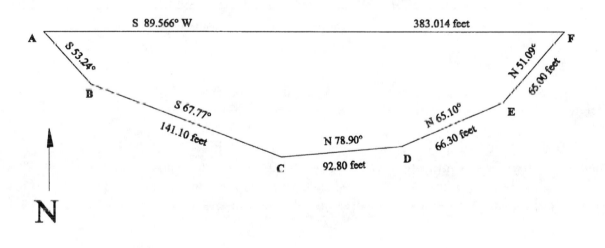

Figure 12-4 Condition 3

Course		Distance		Latitudes		Departures	
No.	Bearing	Cos	Sin	North	South	East	West
AB	S 53.244° E	50	80		30.398	40.700	
BC	S 67.772° E	141	10		53.376	130.614	
CD	N 78.900° E	92	80	17.866		91.063	
DE	N 65.105° E	66	30	27.908		60.139	
EF	N 51.094° E	65	00	40.822		50.581	
FA	S 89.566° W	383	014		2.822		373.097
				86.596	86.596	373.097	373.097

Figure 12-5 Calculations for latitudes and departures

Angle *C* is the back bearing of *BC* plus the NE bearing of *CD*: 67.770 + 78.900 = 146.670.

At *D* the portion of quadrant II that lies within angle *D* is 90°− 81° = 9°. The portion of quadrant II that lies within angle *D* is 90° − 78.90° = 11.10° Thus, angle *D* equals 9° + 11.100 = 20.100. You now know the measurements of three angles and one side. You just need to find two angles and a side to solve this oblique triangle by the law of sines.

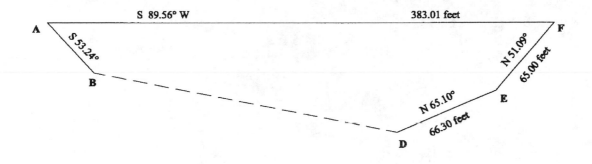

Figure 12-6 *A line from B to D*

Course		Distance		Latitudes		Departures	
No.	Bearing	Cos	Sin	North	South	East	West
AB	S 53.244° E	50	80		30.398	40.700	
BC	S 67.772° E	141	10		53.376	130.614	
CD	N 78.900° E	92	80	17.866		91.063	
DE	N 65.105° E	00	30	27.908		00.139	
EF	N 51.094° E	65	00	40.822		50.581	
FA	S 89.566° W	383	014		2.822		373.097
				68.730	33.220	151.420	373.097
				33.220	35.510	221.677	151.420
				35.510	68.730	373.097	221.677

Figure 12-7 *Calculations for BD*

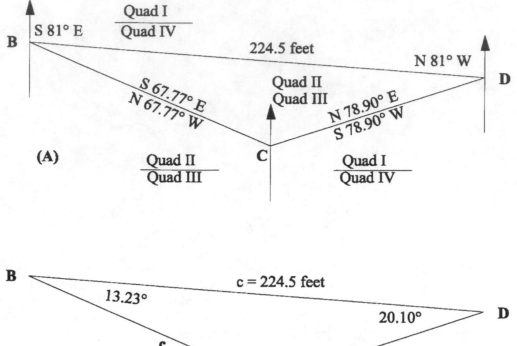

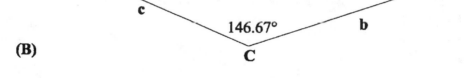

Figure 12-8 The values already known

Use the following calculations to find side *b* of Figure 12-8B:

$b/\sin B = c/\sin C$;

$b/\sin 13.23° = 224.5/\sin 146.67°$;

$b/\sin 13.230 = 224.5/\cos 56.67°$;

$b = 224.5 \times \sin 13.23°/\cos 56.67°$; $b = 224.5 \times .22886/ .54946 = 93.5$ feet

the length of *CD*;

$\sin 146.67° = \cos 56.67°$.

This gives you a difference of 93.5 − 92.8 = 0.7 foot between the value shown in Figure 12-5 and the value you computed by the law of sines. The difference is due to rounding off decimal degrees and decimal footage around the survey.

Use the following calculations to find side d in Figure 12-8B:

$d/\sin D = c/\sin C$; $d/\sin 20.10°$

$= 224.5/\cos 56.67°$;

$d = 224.5 \times .34366/.54946 = 140.41$ feet.

Again you have a difference of $141.10 - 140.41 = 0.69$ foot due to rounding off.

To close the survey mathematically you need to balance the latitude and departure columns and adjust the lengths of the lines as in Figure 12-9. The error of closure is $\sqrt{(0.395) + (0.049)} = .158426 = 0.16$.

The precision is $0.16/799.024 = 1/4994$ (divide 799.024 by 0.16).

An error of 1 foot in 5000 feet (roughly 1 mile) is acceptable for a small survey (0.16 is less than two tenths of a foot for the total survey distance), but in some cases you will want to correct for errors (briefly mentioned in Chapter 3). The following example shows you how to correct latitudes and departures. These corrections are distributed proportionally among all courses of the survey. Consider latitudes first.

To apply a correction to the latitude of each course use this formula: total error is latitude × latitude of the course/sum of the latitudes of all courses. In this case: $0.395 \times 30.398/173.067 = 0.069$ for course AB. And,

Course		Distance		Latitudes		Departures	
No.	Bearing	Cos	Sin	North	South	East	West
AB	S 53.244° E 53.181°	~~50~~ 50	~~80~~ 838		30.398 + .069 30.467	40.700 - .002 40.698	
BC	S 67.772° E	140 .37829	410 .92568		53.116 + .121 53.237	129.975 - .008 129.967	
CD	N 78.900° E	93 .19252	500 .98129	18.001 - .041 17.960		91.751 - .006 91.745	
DE	N 65.105° E	66	300	27.908 - .063 27.845		60.139 - .004 60.135	
EF	N 57.094° E	65	000	40.822 .093 40.729		50.581 - .003 50.578	
FA	S 89.566° W	383	014		2.822 + .006 2.828		373.097 + .024 373.121
		799	024	86.731 86.336 0.395	86.336 0.395 86.731	373.146 373.097 0.049	373.097 0.049 373.146
				173.067		746.243	

Figure 12-9 *Balancing a survey*

Course $BC = 0.395 \times 53.116/173.067$ 0.121,

Course $CD = 0.395 \times 18.001/173.067$ 0.041,

Course $DE = 0.395 \times 27.908/171.067 = $ 0.063,

Course $EF = 0.395 \times 40.822/173.067 = 0.093$, and

Course $FA = 0.395 \times 2.822/173.067 = 0.006$.

Since the North total is the larger (86.731) of the two columns, subtract each correction as shown under each course latitude to get a new total of 86.532. The difference between column totals now is $86.534 - 86.532 = 0.002$ feet, a negligible quantity.

Similarly, compute the corrections for the departures:

Course $AB = 0.049 \times 40.700/746.243 = 0.002$,

Course BC = 0.049 × 129.975/746.243 = 0.008,

Course CD = 0.049 × 91.751/746.243 = 0.006,

Course DE = 0.049 × 60.139/746.243 = 0.004,

Course EF = 0.049 × 50.581/746.243 = 0.003, and

Course FA = 0.049 × 373.097/746.243 = 0.024.

Since the East column is the larger (373.146) of the two columns, subtract each correction as shown under each course to get a new total of 373.123. The difference between column totals now is 373.123 − 373.121 = 0.002 feet, a negligible quantity. Add the corrections to each West course.

Since you applied corrections to latitudes and departures, you also need to correct the lengths and bearings of the courses.

For course AB, the length = $\sqrt{(30.467)2 + (40.698)2} = 50.838$

Mark this in as a new length and draw a light line through the former length. Whenever you are surveying, always draw through wrong information rather than erasing. The time may come when the information is important.

For course AB, the tan of the bearing = 40.698/30.467 = 1.33580 and the arc tan 1.33580 = 53.181°. Mark this into the column under Bearing and draw a light line through the former bearing.

Compare the two bearings: 53.244° = 53°14'38" and 53.181° = 53°10'51". The difference is 00°03'47". If you had turned the new bearing on the previous course, the difference would be slightly over one tenth of a foot.

For practice, work out the remainder of the courses, convert them to degrees, minutes, and seconds, and make a comparison similar to the one above.

■ Solving for Condition 4

Condition 4, in which the bearings of two lines are missing from the survey given in Figure 12-4, is shown by Figure 12-10A. These are the same lines, BC and BD, that you found the lengths for in Figure 12-6.

This time, use the same procedure to find a triangle with three lengths given and two bearings missing, as shown in Figure 12-10B.

Draw a line from B to D and find its length and bearing as in Figure 12-7. Then, form the triangle BCD for which you know the lengths of three sides: BC = 141.10 feet; CD = 92.80

feet; BD = 224.50 feet. Solve the triangle BCD for the angles using the law of cosines and the missing bearings you computed from the angles.

Following conventional practice, identify the side opposite each angle with a lowercase letter: angle B, side b; angle C, side c; angle D, side d. Then, use the law of cosines which says that when three sides are given:

$$\cos B = (c^2 + d^2 - b^2)/2cd,$$

$$\cos B = [(224.50)^2 + (141.10)^2 - (92.80)^2]/2cd$$

$$= 61697.62/63353.90 = .9738567,$$

and arc cos .9738567 = B = 13.13°.

$$\cos C = (b^2 + d^2 - c^2)/2bd$$

$$= [(92.80)^2 + (141.10)^2 - (224.50)^2]/2 \times 92.80 \times 141.10$$

$$= -21879.20/26188.16 = -.835461522,$$

and arc cos − .835461522 = C = 146.664°.

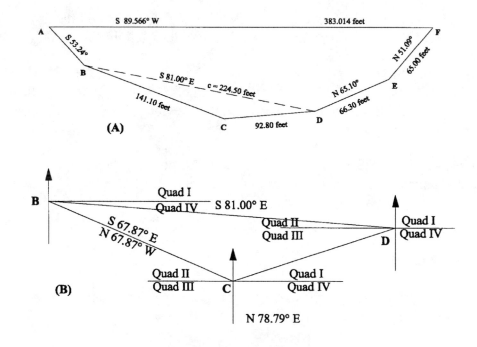

Figure 12-10 Condition 4

$$\cos D = (b^2 + c^2 - d^2)/2bc$$

$$= [(92.80)^2 + (224.50)^2 - (141.10)^2]/2 \times 92.80 \times 224.50$$

= 39102.88/41667.20 = .938457,

and arc cos .938457 = D = 20.206°.

To proof your work, check to see that 13.13° + 146.664° + 20.206° = 180°, the sum of the degrees in the three angles of a triangle. The minus sign before cos C (−.835461522) indicates that angle C is greater than 90° but less than 180°.

Another way you can find angle C is to find angles B and D and subtract their sum from 180°: 180° − (13.13° + 20.206°) − 146.664°.

The triangle BCD is shown in Figure 12-10B with the bearing S81.00°E. You could find this bearing in Figures 12-6 and 12-7 by drawing the line BD to eliminate lines BC and CD. As shown, indicate the quadrants to orient the line with respect to the bearing. Both the bearing S81.00°E and line BC lie in quadrant IV.

Since the bearing is referenced from the south direction, 81.00° − angle B = 81.00° − 13.13° = 67.87°. Line BC lies on the bearing S67.87°E.

At angle C use the northwest bearing (the back bearing) of line BC to determine the portion of angle C that lies in quadrant II: 67.87°. Then, 146.664° − 67.87° = 78.794°, the bearing of line CD. Angle C is the sum of the angles that lie on each side of the north sign, one in quadrant I and one in quadrant II.

At angle D, if you use the back bearing N81.00°W, the remainder of the 90° in quadrant II is 90° − 81° = 9°. This is part of angle D. The remainder of the 90° in quadrant III is 90° − 78.79° − 11.21°. This is part of angle D.

Then, 9° + 11.21° = 20.21° (20.206° as found before). Note that the small differences in bearings and angles are from previous calculations due to rounding off the digits' display in a calculator.

■ Other Conditions

All problems of closure and missing measurements involve the mathematical relationships of bearings and lengths, and latitudes and departures. You have already learned how to evaluate some of the most common situations. For other missing measurements, use the formulas listed in Table 12-1 with the techniques you already know.

Table 12-1. Missing Measurement Formulas

Missing Information	Information You Have	Formula
Latitude	Bearing and Length	Length × cosine bearing
Departure	Bearing and Length	Length × sine bearing
Length	Latitude and bearing	Latitude/cosine bearing
Length	Departure and bearing	Departure/sin bearing
Tangent of bearing	Latitude and departure	Departure/latitude
Cosine bearing	Latitude and departure	Departure/latitude
Sine bearing	Departure and length	Departure/length
Length	Latitude and departure	Length = $\sqrt{(\text{latitude})^2 + (\text{departure})^2}$

Subdividing Land

The formulas you have used to find missing measurements come in handy when you need to measure the size of lots of land in a subdivision. Often the owner will already have marked off points on the land that establish the survey courses. In such cases you just need to find the lengths, bearings, and area in order to write the legal description. Other cases are more difficult.

For instance, although lots in subdivisions are usually rectangular, they sometimes have irregular sizes and shapes. The shapes may depend on the lay of the land and the sizes of the planned buildings. You may be asked to measure an irregular lot that is already staked out, or to create an irregular lot according to certain specifications.

The following example shows you how to approach complex subdivision situations.

■ Example: Dividing the Farm

Suppose a farmer's will specifies that his farm be divided equally among three sons. The farm has 15 sides (survey courses).

The first thing to do is have a conference with the three sons. They were brought up on the farm and probably have an idea of how they'd like to divide it. They may even have chosen certain corners or points on the farm boundaries that almost divide it naturally into thirds.

Next, survey the farm and draw it to the largest practical scale. Then, start dividing the land.

To begin, choose two points (possibly those suggested by the sons) and draw a line that seems to mark off a one-third portion. Close this figure using the methods given in Figure 12-6 and 12-7. Then compute the area to see how close it is to being a one-third portion.

Assume that the closing line between the points has a bearing of N20'30'W and is 1100 feet long. You find that the area is 3 acres less than one third of the total acreage. Your problem now is to define exactly 3 acres to add to the tract.

Figure 12-11 shows the closing line between two corner points and the bearings (N83'40'E and S76'15'E) of the continuing boundary lines of the farm. If you draw a line between these two lines and parallel to the closing line, you get a trapezoid. You can size the trapezoid to contain the 3 acres you need.

Note that the closing line and the broken line parallel to it are the bases b and b' of the trapezoid. The broken line begins at the boundary line S76'15'E, continues through boundary line N83'40'E, and ends on a line at 90° to the closing line. Both lines designated as H are at 90° to the closing line. Subtract the part of b' (3.03 feet) that is above line N83'40'E from the line as shown in Figure 12-11A. Add the part of b' (8.17') that is below line H to b' to complete the b' base of the trapezoid.

Choose a trial altitude (H) for the rectangle that lies between the lines designated as H. For example, try 100 feet. Then, $100 \times 1100 = 110,000$ square feet and $110,000/43,560 = 2.525$ acres. This is about 1/2 acre short of 3 acres.

Since you get 5 1/2-acre parcels in 2.525 acres when you use an altitude of 100 feet, each 1/2-acre comes from 20 feet of altitude ($100/5 = 20$). To get the 1/2 acre you still need, add 20 feet to the altitude.

When you do this, $120 \times 1100 = 132,000$ square feet and $132,000/43,560 = 3.03$ acres. For practical purposes this is close enough to the 3 acres you wanted. Note that $43,560 \times 0.03 = 1306.8$ square feet, and $\sqrt{1306.8} = 36$. The area you parted off is an equivalent square with 36-foot sides.

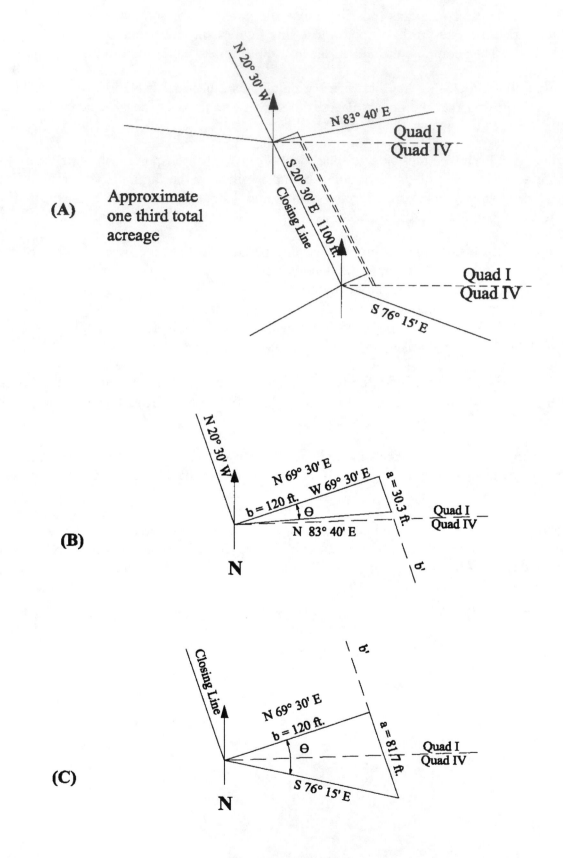

(A) Approximate one third total acreage

N 20° 30' W

N 83° 40' E

Quad I
Quad IV

S 20° 30' E 1100 ft.

Closing Line

Quad I
Quad IV

S 76° 15' E

(B)

N 20° 30' W

N 69° 30' E

W 69° 30' E

b = 120 ft.

a = 30.3 ft.

Θ

N 83° 40' E

Quad I
Quad IV

b'

N

(C)

Closing Line

N 69° 30' E

b = 120 ft.

Θ

a = 81.7 ft.

Quad I
Quad IV

b'

S 76° 15' E

N

Figure 12-11 The closing line

In Figure 12-11B, the area within the triangle lies outside of the property and you need to subtract it. In Figure 12-11C, the area within the triangle lies outside the rectangle bounded by lines H. Add it so that it becomes part of the trapezoid within the property. Altitude H (b = 120 feet), in Figures 12-11B and 12-11C is at 90° to the closing line (bearing N20'30'W).

Next, consider the triangle seen in Figure 12-11B. The angle between the north sign and the triangle's base b in quadrant I is

$$90° - (20°30') = 69°30'.$$

And,

$$\theta = (83°40') - (69°30') = 14°10',$$

$$\tan \theta = a/b,$$

$$a = b \times \tan \theta = 120 \times .25242 = 30.3 \text{ feet.}$$

The area of the triangle is $ab/2 = (30.0 \times 120)/2 = 1818$ square feet. This is equivalent to a 42.6-foot square area.

Now, consider the triangle seen in Figure 12-11C. The altitude H is at 90° to the closing line. This gives you the bearing for line H as it did before: N69°30'E. The abscissa (the broken line) divides the triangle between quadrant I and quadrant IV. The part of the triangle that lies in quadrant I is $90° - (69°30') = 20°30'$. The part of the triangle that lies in quadrant IV is $90° - (76°15') = 13°45'$. The total angle, θ, is

$$(20°30') + (13°45') = 34°15'.$$

$$\tan \theta = a/b,$$

$$a = b \times \tan \theta$$

$$= 120 \times .68087 = 81.70 \text{ feet.}$$

The area of the triangle is $ab/2 = (81.70 \times 120)/2 = 9804/2 = 4902$ square feet. This is equivalent to a 70-foot square area.

Next, find a line b' that extends from property line N83°40'E to property line S76°15'E. The long sides of the rectangle between the two short sides (altitude H) are 1100 feet long.

Subtract a = 30.3 feet from the data in Figure 12-11B: 1100 − 30.3 = 1069.7 feet. Add a = 81.7 feet to the data in Figure 12-11C: 1069.7 + 81.7 = 1151.40 feet. The area of the trapezoid is $1/2 \ a(b + b')$ and $1/2 \times 120(1100 + 1151.40) = 60 \times 2251.40 = 135,084$ square feet. If you subtract the area of the triangle in 12-11B, you have 135,084 − 1818 = 133,266 square feet.

Adding the area of the triangle in Figure 12-11C you get 133,266 + 4902 = 138,168 square feet. The area within the trapezoid is 138,168/43,560 = 3.17 acres. The average is 0.17 acre, more than 1/4 acre, or 43,560 × 0.17 = 7405.2 square feet. This is equivalent to a square area with 86.05-foot sides.

You hope that the brothers will be satisfied with this condition. But, if they want a closer approximation, you are prepared to adjust the area.

Consider a 10-foot wide strip (b'') along side b'. The area between b' and b'' is 10 × 1151.40 = 11,514 square feet. This is not a large enough area. Therefore, put a few more trial figures for altitude into a calculator. Try 10.2 feet and 10.4 feet. Finally, 10.5 × 1151.40 = 12,193.33 square feet and 12,196.8−12,193.33 = 3.47 square feet. So, reduce the altitude by 10.5 feet.

This last calculation assumes that the area is a rectangle, which it is not (due to the bearings of the two continuing property lines). But for practical purposes the 120-foot altitude can be reduced as stated above. And, with careful work you can compute the area with precision.

You now need to add the final closing line. Set a stake on each line H at a distance equal to the altitude of the trapezoid, set up on one of the stakes, and sight in the other stake and set points on each property line to mark the closing line.

Follow the same steps to divide the remaining two thirds.

CHAPTER 13

Mapping The Site

After you have done all of the field work for a survey, you need to draw a map of the site. In a sense, mapping a site is like doing the field work-measuring distances, locating sewer lines, setting elevations, and so on except, this time, on paper. Accuracy is as important in mapping as it is in taking the field measurements.

Mapping Procedure

■ Order of Mapping

Do your mapping in the same order as you did the field work. That is, start with the same point first set in the field, set it first on the map, do the same with the second measurement taken, and continue adding information to the map in the order it was discovered in the field.

■ Distances

In the field, you measure distances with a tape. In the drawing room, you measure distances with a scale. You have a choice of scales, depending on the size of the site. For instance, if you were to use a 10-foot per inch scale to draw a large site, the drawing would be outsized. An engineer's scale is 12 inches long and for a 10-foot per inch scale a 12-inch scale measures only $10 \times 12 = 120$ feet.

For most mapping you will do best to use a 50-foot per inch scale. In any case, start by taking the longest distance on the site and laying it out with a 50 scale. This scale gives you 600 feet for 12 inches of scale length. If you divide the longest site distance by 600, you will find the length of drawing paper you will need.

If you measure carefully and keep the multiple of 10 in mind, the 50 scale can become the 500 scale. There are 50 divisions per inch on the 50 scale; each division represents 1 foot. If you use it as a 500 scale, each division represents 10 feet. You could also use it as a 5000-foot per inch scale, each division representing 100 feet.

The 60 scale has 60 divisions per inch and could be used as either a 600-foot or a 6000-foot scale. And, of course, these methods apply to any size scale you use. If you practice using scales to dimensions other than those they actually show, you will be able to draw unlimited scale distances.

■ Angles

In the field you use a transit to measure angles. On paper, use a protractor instead. You can plot very accurately using the tangent, sine, and cosine methods discussed in Chapter 14.

The north direction is universally accepted as being toward the top of the map. The side lines on the map run true north. The top and bottom lines run true east and west. You can quickly transfer these directions to any point for plotting angles (see Chapter 14).

■ Using Multiple Maps

Sometimes you will want to draw several maps rather than placing all of the information on one map. On the first map, show the "as is" area. Indicate the land features to be changed, how the changes are to be made, and which features are to be completely removed from the site (boulders, bridges, barns, and so on). Show a creek to be rerouted, a swamp area to be filled, and trees to be removed. Put it all on the first map.

On the second map, show the area as it is intended to be after development. Show all changes as made: creek and bridges changed; lots laid out; streets, sidewalks, approach roads shown and staked; utility lines, sewers, catch basins and manholes as they'll be installed; fire hydrants; trees; surveying data; contours; building elevations (sometimes basement and each floor); sections through streets; driveways; the established limits of creek flood plains; and so on.

Show design technology on separate sheets. This includes construction details such as on-site waste-water treatment facilities, yard drains and curb inlets, sewers, storm drains, catch basins, and manholes.

General Points

The maps you draw should be neat and accurately drawn. Use a pencil for the first sketch, and then go over the pencil lines with ink. Draw the borders last.

Reserve a space (generally in the lower right-hand corner) for a title block. Show a short, explicit title; the names of the surveyors, engineers, drafters, and field crew; and the scale of map and date.

Other items you should include wherever convenient are the north sign to true north, a legend of the symbols used to depict features of the map, and explanatory notes. Group general notes in one space, usually along the right side above the title block. You can put small specific notes near the feature to which the information is related.

Show boundary lines and boundary-line intersections referenced to some permanent object. Show monuments, fences, names of streets, streams, adjoining property owners (with the number of deed book and page), and, if it is a rural development, give the direction and distance to the nearest town. Show the value of the site's corner angles. Show a distance and bearing from a site corner to some legal point such as a township corner or section corner.

Be sure to give the acreage of each lot, street, playground, alley, and other areas shown. The county recorder and county engineer need this information for several purposes.

Draw curves first and then add straight lines drawn tangent to them at the point of contact.

Draw contours clearly and carefully note the elevations and other pertinent data nearby.

Place lettering, notes, dimensions, and angles, so that they can be read either from the bottom or the right side of the map. Place the name and acreage of streets, lots, streams, or other areas within the respective areas on the map. This information is important to county officials and the taxpayers of the development. You can put general information, such as the names of adjoining property owners, in any clear space in the general area of the property so designated.

You can buy many lettering devices at engineers' supply houses. The styles of lettering are sometimes regional and you can find them on maps in the county engineer's or county recorder's office.

Before Construction Begins

Be certain that every map and drawing is approved for construction and that all legal work is completed before the first small bit of earth is turned to start construction. You will find that thousands of dollars will be wasted if you need to correct a layout or redo a building that was not approved before construction (and even sales) began.

During construction, check proposed changes against codes, ordinances, rules, and regulations. If you provide careful supervision and planning of all changes and keep a record by mapping you can prevent construction, municipal, and legal problems.

Finally, no construction will completely follow your original plans. As construction proceeds, changes will be made. When construction is complete map the changes for a permanent record on "as built" drawings.

CHAPTER 14

Plotting Angles

When you are in the field, especially when you are mapping a survey, it is important to know that you have a choice of four methods to plot angles: (1) using a protractor; (2) by trigonometric methods, using the tangent of an angle or the sine or cosine of an angle; (3) by the chord of an angle; and (4) by latitudes and departures or coordinates.

So far, the angles you have used in this book have been presented as lying within the vector circle, or four-quadrant figure. The horizontal line of that figure is called the *abscissa*, and the vertical line is the *ordinate*. The lines intersect perpendicularly to form four 90° angles.

Using a Protractor

To plot an angle using a protractor, place the vertex of the angle at the intersection point of the abscissa and ordinate. Place a dot with a sharp pencil outside the protractor and in line with the angle you want.

Remove the protractor and draw a line from the vertex through the dot to whatever length you choose.

Using Trigonometry

■ Tangents

In Figure 14-1 a 30° angle and a 33° angle are laid off. Since it is a 90° angle from the zero mark around the arc of the protractor to the horizontal line, marking 60° from zero gives you a 30° angle from the horizontal line. Marking 57° from the zero mark gives you a 33° angle from the horizontal line.

In any right triangle the length of the base times the tangent of the toe angle gives you the altitude of the triangle. If you assume the base is 300 feet long and a 28° angle is required at the vertex, then, the altitude is:

$300 \times \tan 28° = 300 \times .53171 = 159.5$ feet.

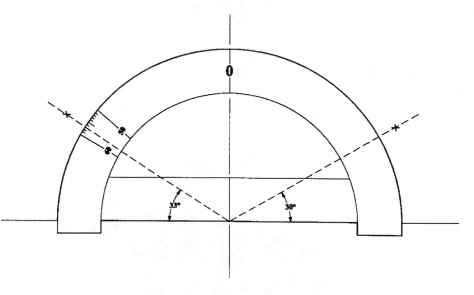

Figure 14-1 *Finding the angle from the horizontal line*

Draw a perpendicular from the 300-foot mark as shown in Figure 14-2. Place a mark at 159.5 feet on the perpendicular (shown by the cross) then draw a line from the vertex through the mark. The vertex angle is 28°.

If the angle is somewhat larger than 45°, the tangent (altitude) becomes very long. In this case, it is best if you plot the complement of the angle (Figure 14-2B).

Assume the base is 300 feet and the angle required is 60°. If you draw a perpendicular (broken line, Figure 14-2B), at the horizontal 300-foot length, you will need considerable space to erect a perpendicular long enough to establish the angle. Laying this out in the field could cause you numerous problems. But, if you plot the complement of the angle (30°), the altitude of the angle is easy to find. Thus, altitude is

$$300 \tan 30° = 300 \times .57735 = 173.2 \text{ feet.}$$

Establish a point on a line at 90° off the vertical at the 300-foot mark and 173.2 feet from the vertical line (at the cross). A line from the vertex of the angle through this point is one side of a 60° angle.

■ Sine and Cosine

The sine and cosine methods are similar to the tangent method. To plot 28°, use sine = 0.469 and cosine = 0.883. Multiply these by 10 and lay them out at 90° as inches (or any increment of measure). The cosine is the base at 8.83 inches and the sine is the altitude at 4.69 inches.

The hypotenuse should equal $\sqrt{(8.83)^2 + (4.69)^2} = 9.99,$ about 10 inches (see Figure 14-3). If the angle is somewhat larger than 45°, plot the complement or supplement as shown using the tangent method.

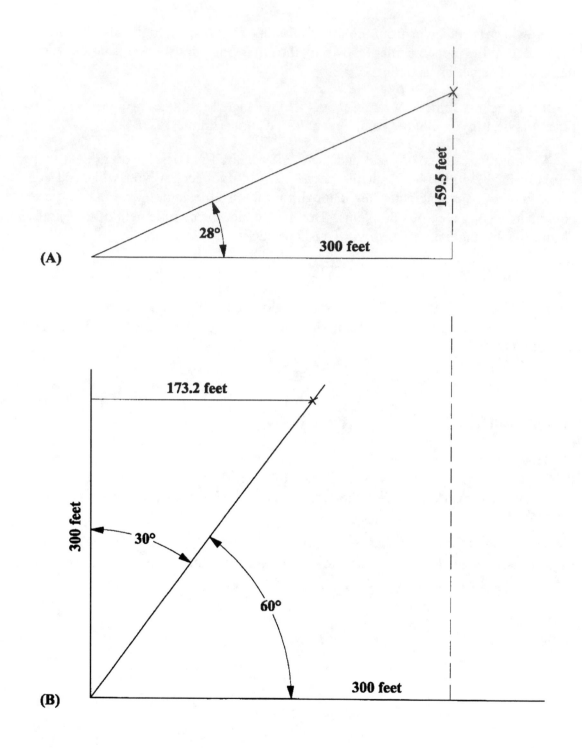

(A)

159.5 feet

28°

300 feet

(B)

173.2 feet

300 feet

30°

60°

300 feet

Figure 14-2 *A perpendicular from the 300-foot mark*

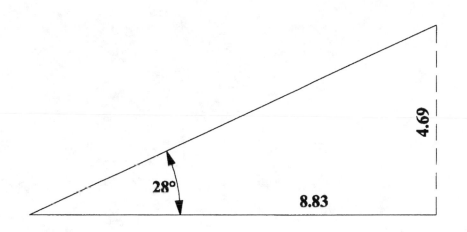

Figure 14-3 *Sine and cosine methods*

■ Using the Chord Method

You can also use the chord method by constructing the angle with two intersecting arcs. To do this plot an angle of 30°10'. Take a point O on a baseline as shown in Figure 14-4. With some whole-number dimension, say 10 inches, strike the first arc (*AB*) using point O as a center.

The radius of the second arc must equal the chord of the given angle that corresponds to the radius of the first arc (the dotted line from point *D*). Use a chord table to find out that the chord for 30°10' for a unit (1) radius is 0.5204 inches. For 10 inches the chord is 5.205 inches (multiply by number of inches).

Using 5.204 inches as a radius, draw an arc (*EF*) with point *D* as a center to intersect arc *AB*. Draw a line from point O through this point of intersection to establish the angle 30°10'.

Many chord tables, however, do not list every angle and fraction of an angle. If a chord-table value is not available, find the sine of one half of the required angle and multiply this value by twice the length of the radius of the first arc. One half of 30°10' = 15°05'. The sine of 15°05' is .2602 to four decimal places. Twice the length of the radius of the first arc is $2 \times 10 = 20$ inches and $20 \times .2602 = 5.204$ inches, the same value used for the radius of the second arc.

For angles larger than about 73°, draw a perpendicular from the vertex point and plot the complement of the required angle.

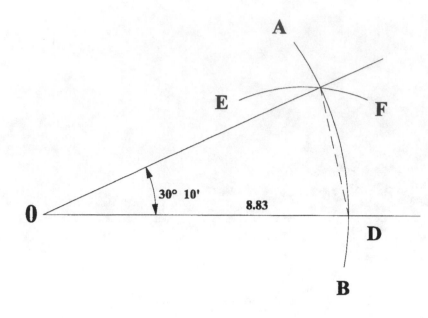

Figure 14-4 *The chord method*

■ Using Latitudes and Departures

The latitude and departure method is not generally used for laying out a single angle. But if a figure has several sides, such as when you are doing a survey, this method is very useful. You can see how this method works in Figures 12-1 and 12-2 where it was used to find the bearing of line *BC*. You get the tangent of the bearing by dividing the departure by the latitude. The arc tan of the tangent gives you the angle.

CHAPTER 15

True North, Latitude, and Longitude

Although most surveyors use magnetic north (found with a compass) as their reference line, this book uses a true north meridian instead. Since a true north meridian is located and identified by celestial observation, it is more precise than magnetic measurement. This chapter gives you a procedure for finding true north and also describes how to locate a point by latitude and longitude.

Locating True North

To find a true north line you can observe Polaris (the north star), which moves in a circular pattern with respect to the north pole. Twice daily Polaris is on a line (meridian) that lies between you (the observer) and the north pole, once at upper culmination and once at lower culmination. Upper culmination is when the star is at a point on its circle above and closest to you. Lower culmination is when the star is on its circle at a point nearer the horizon and away from you. You find a true north meridian by observing Polaris at either point of culmination. The line between you and the point of culmination is a true north meridian. You can look up the times of culmination for Polaris in a current ephemeris (a celestial calendar).

■ The Equal Altitude Method

The method you use to sight Polaris is the *Equal Altitude Method* and, since the Big Dipper appears to rotate around Polaris, you can actually use any star in the Big Dipper for your observation. As described for Polaris, during one complete revolution of the celestial sphere, these stars cross the meridian twice. As the star you have chosen passes above you (at the *observer's zenith*) it reaches its highest altitude at a point on a meridian between you and the north pole (the upper culmination).

It then travels westward and descends below the horizon southwest of north to pass behind the pole. As it crosses a projection of the same meridian, it is at its lower culmination. The star then continues to rotate and rises in the east.

Try using the end star of the Big Dipper handle. It is easy to find just after sunset and just before sunrise. You can fix a line of direction on it at a certain altitude as it descends westerly and then again at the same height as it ascends in the east. You want to fix a line of direction on the star at a certain altitude as it descends and measure a horizontal angle from a fixed point. As the star rises the next morning in the east, you will measure the horizontal angle from the star at the same altitude to the same fixed point.

Taking Sightings. Follow the method in Figure 15-1 to prepare station 0 (stake and tack) shielded from night lighting and any possible local magnetic attraction (cars, fences, or any metal objects). Set station 1 (stake and tack) a few hundred feet north of and visible from station 0. You can use a point on some prominent feature of the landscape for this station.

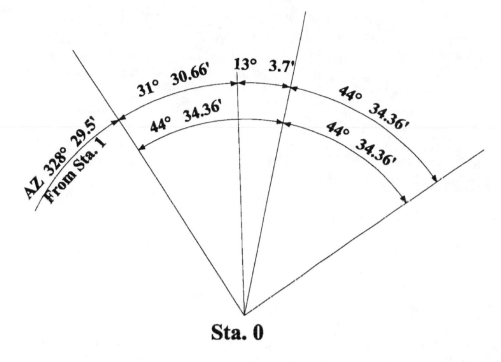

Sta. 0

Figure 15-1 *Preparing stations*

At about the time you expect the stars to appear (for example, 8 PM), set up the transit on station 0, clamp the plates at zero, and level the transit. Sight on station 1. If it is dark, place a light beam on station 1 and on the telescope crosshairs. A flashlight beam directed from a side angle to the crosshairs works well.

Find your star, the end star of the Big Dipper in this example, which is descending southwest of north. Loosen the top plate and sight on the star. Clamp the vertical arc on the nearest even degree below the star.

Clamp the top plate and use the tangent screw and vertical hair to follow the star as it descends. When the star centers on the cross hairs, record the time, vertical angle, and horizontal angle. The vertical angle is the altitude and the horizontal angle is an azimuth angle turned from station I (turned right from station 1 as AZ 328°29.5' in Figure 15-1. This is the averaged angle).

Again, loosen the vertical arc, set it at the next lower half-degree or degree, and follow the star to the cross hairs. Record the same values. Repeat this process for the next half-degree or degree. Now you have made three observations during the evening period.

Since these observations were begun at 8:00 PM, you will set the transit over station 0 at about 3:30 AM to repeat the same observations on the same star at the same altitudes and measure three horizontal angles from the eastern ascension of the star.

Table 15-1 shows the notes for the same procedure performed September 1, 1985. You can model your notes and calculations after this chart. The latest evening time shows the lowest (descending) vertical angle. The earliest morning time shows the lowest (ascending) vertical angle. The morning time is inverted in the table to place relative altitudes on the same horizontal line.

The midpoint of each set of vertical readings (evening and morning) is calculated then averaged. In column 1,

$$360 \ (359–60) – (328–12) = 31–48.$$

$$(31–48) + (56–37.5) = 87–85.5 = 88–25.5.$$

$$(88–25.5)/2 = 44–12.75. \quad (56–37.50) –(44–12.75) = 12–24.75,$$

the midpoint between the lines of direction taken to the star evening and morning. Follow the calculations through the averaged results (Table 15-1).

Set up on station 0, clamp the plates, and sight on station 1. Loosen the top plate and turn the averaged angle (13°3.7') from station 1. Monument a point on this line: that point is true north.

When surveying, you should always have a true north line for checking the angle of declination. If you are doing a re-survey, and the original survey was referenced from magnetic north, the difference between the magnetic declination at the time it was originally surveyed and today can be determined.

Table 15-1. Observation Notes

Time	Vertical ∠	Horizontal ∠	Time	Vertical ∠	Horizontal ∠
1. 8:16 PM	28°06'	328°12'	4:27 AM	28°06'	56°37.5'
2. 8:20 PM	27°36'	328°28.5'	4:22 AM	27°36'	56°21'
3. 8:24 PM	27°06'	328°47.5'	4:16 AM	27°06'	55°58'

1. 359-60	2. 359-60.0	3. 359-60.0
− 328-12	− 328-28.5	− 328-47.5
= 31-48.0	= 31-31.5	= 31-12.5
+ 56.37.5	+ 56.21.0	+ 55.58.0
= 87.85.5	= 87.52.5	= 86-70.5
= 88.25.5	÷ 2	÷ 2
÷ 2	= 43.5-26.25	= 43-35.25
= 44-12.75	= 43-56.25	& 55-58.00
& 56-37.50	& 56-21	− 43-35.25
− 44-12.75	= 55-81.00	= 12-22.75
= 12-24.75	− 43-56.25	midpoint from station 1
midpoint from station 1	= 12-24.75	
	midpoint from station 1	

< AVERAGED >

+ 12-24.75	+ 328-12.0	+ 56-37.5
+ 12-24.75	+ 328-28.5	+ 56-21.0
+ 12-22.75	+ 328-47.5	+ 55-58.0
= 36-72.25	= 984-88.0	= 167-116.5
÷ 3	÷ 3	÷ 3
= 12-24.083	= 328°29.5	= 55.66-38.83
& 12-24.083	average Horizontal ∠	= 56°18.43'
= 12-24.083 answer		average Horizontal ∠
midpoint station 1		

September 1, 1985. Station 0 on stake and tack referenced from the NE and NW corners of surveyor's home and shown on the house plans in vellum file. Station 1 on 2-inch pipe in concrete base on surveyor's farm north of station 0. Readings were taken on the end star of the Big Dipper handle.

Locating By Latitude and Longitude

You will often want to find a point by latitude and longitude. To do this, use a quadrangle map. Figures 15-2 and 15-3 show that any point can be located on such a map. The map shown is taken from the 1929 edition of the West Virginia-Ohio Point Pleasant quadrangle map. Notice the point shown as a small rectangle mid-river near Gallipolis, Ohio. The point represents a boat.

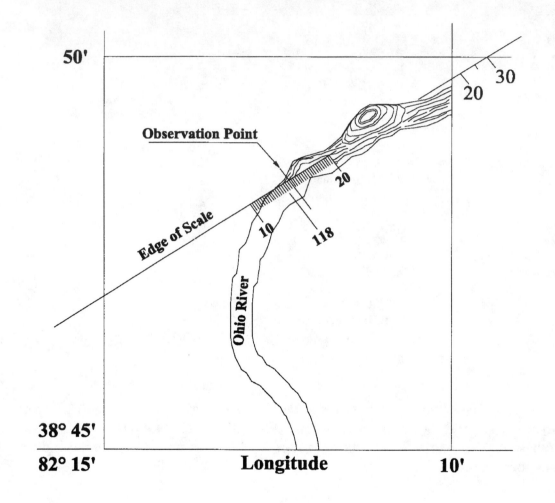

Figure 15-2 A quadrangle map

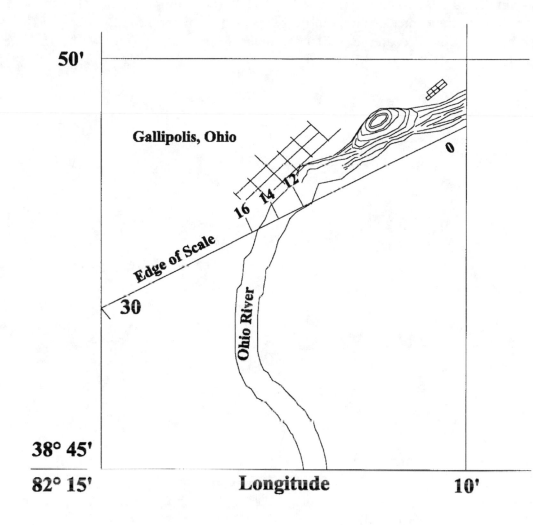

Figure 15-3 *Finding the latitude and longitude of a point*

To locate the boat by latitude and longitude place the map over a sheet of drawing paper and extend the lines you will need in order to place a scale ruler.

In Figure 15-2 the lower line is extended so that zero on the 30 scale lies on this line, 30 (here 300) on the scale lies on the 50 minute line, and the edge of the scale crosses through the center of the point to be located. The scale reading at the center of the point to be located is 188, which is the number of seconds to be added to the latitude at the bottom. Converted to minutes, 188/60 = 03'08". The altitude of the point then is (38°45') + (03'08") = 38°48'08".

The scale interval zero to 300 is used because 300/60 = 5 minute distance from 38°45' to 38°50'.

You can find the longitude using the same method by spanning the space between 82°10' to 38°15'. Again, this is a span of 5'.

Place zero on the 30 scale on the 10 minute line and 30 (here 300) on the 82°15' line with the scale edge through the center of the point to be located. This point is found at 14 (here 140), the number of seconds to add to 82°10'. Convert to minutes, 140/60 = 02'20". The longitude then is (82°10') + (02'20") = 82°12'20".

APPENDIX

A.

The Transit

B.

Geometry for Construction

C.

Trigonometry for Construction

D.

More Practical Examples

APPENDIX A

The Transit

This section is a review of how to set up a transit, turn angles, and take readings.

Using a Transit

Generally, a builder's transit has a horizontal circle graduated in intervals of 1° and continuously numbered around the circle in units of 10°. Some instruments have double numbering, which runs both clockwise and counterclockwise.

To measure an angle, set up the instrument over a reference point - that is, the apex of the angle - set the horizontal scales at zero-to-zero degrees, and direct the line of sight to the point from which you want to turn the angle.

Clamp the telescope support in place with the thumb screw. Then adjust the line of sight minutely using another thumb screw known as the *tangent* screw.

Loosen the horizontal scale clamp to release the upper scale and revolve the telescope to sight on the object to which you're measuring the angle. Again, clamp the telescope in place, use the tangent screw for precise sighting, and read the approximate angle.

Then, you get the final degree of accuracy by making a vernier reading. The *vernier scale* is a short scale facing the divisions of the circle scale. You use the vernier to read the fractional part of the degree reading, generally to 5 minutes of a degree on a builder's transit or to 20 seconds of a degree on an engineer's transit.

■ Reading the Vernier

Since you use the vernier reading to determine the fractional part of the smallest division of an angle, you'll need a magnifying glass to accurately read the vernier.

Figure A-1 shows a vernier (A), a short scale (B), and the scale reading from 0 through 10 (C) placed adjacent to a graduated scale.

The vernier index is at the zero point on the short scale (these illustrations are drawn to 1/10 scale).

Figure A-1A shows ten 1/10th marks in a 1-inch space on the graduated (upper) scale. The vernier, adjacent below, shows a scale 9/10 inch in length, divided into ten spaces. Each division of the vernier is 9/10 of a scale division.

The zero matches zero and ten on the vernier scale matches nine on the short scale. The one mark on the vernier falls 1/10 of 1/10 second (1/100 second) short of the 1/10 second mark on the scale.

The two mark on the vernier falls 2/10 (2/100 second) short of the 2/10 second mark on the scale. Successively, each mark falls short its own value until the ten mark falls 10/10 (10/100 second) short and matches the nine mark on the scale.

In Figure A-1B the vernier is moved to match the one mark on the vernier with the 1/10 second mark on the scale. The vernier has moved 1/100 second right. If it is moved to match the two mark on the vernier with the 2/10 second mark on the scale, it has moved 2/100 second right.

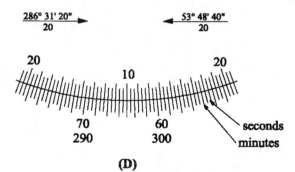

(D)

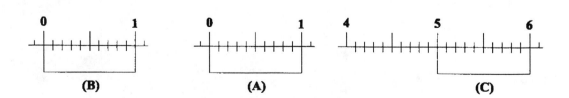

(B)   (A)   (C)

Figure A-1 The vernier

In Figure A-1C a reading is illustrated. The zero mark on the vernier falls just short of 5 on the scale, between 4.9 and 5. The 9 mark on the vernier matches a mark on the scale. This places the zero mark (index) of the vernier 9/100 second (0.09) right of the 4.9 mark on the scale to get a reading of 4.99 seconds. The scale reading is 4.9, the vernier reading 0.09. Adding, 4.9 + 0.09 = 4.99.

Thus, to read a vernier: Take the scale reading nearest the index on the vernier. Find the mark on the vernier that matches a mark on the scale and add it to the scale reading.

Figure A-1D shows a circle and vernier. The circle is divided into 20-minute spaces. The vernier is divided into 20-second spaces. There is an inner reading and an outer reading on the scale.

If you want to read the angle clockwise, set the 20 index at the right end of the vernier to zero on the inner scale.

Often you won't be able to see the degree mark and you must count to the right from the first visible numbered mark (here the mark is 60).

Read from the right toward the left on the inner scale, and you'll see the last numbered degree mark passed by the 20 is 50. Each main division is 1 degree, and 3 degrees have passed by the 20 index on the vernier.

Count right from 60. Six degree marks are counted and 60 − 50 = 10; 10 − 6 = 4. But the 20 mark doesn't match the 4. So, record a reading of 53°.

Then, since each small division on the degree scale equals 20 minutes, and the 20 index has passed two of them, the reading now is 53°40'.

Next, take the vernier reading. Each main division on the vernier scale equals 1 minute and each small division (space) equals 20 seconds.

So, reading from the right-hand 20 mark toward the left, the 40-second mark between the 8- and 9-minute marks is matched to a mark on the scale. The reading is 8 minutes, 40 seconds on the vernier. Add 53°40' and 8'40", to get the total reading of 53°48'40".

Finally, take a counterclockwise reading. Read the outer scale from the left. The left index (20 mark) has passed 286°20'. Carefully check the divisions on the scale to read the outer numbers.

Reading the vernier from the left index (20 mark) toward the right, the 20-second mark beyond the 11-minute mark is matched to a mark on the scale. This is the same mark as in the previous reading. So, this vernier reading is 11'20". Add 286°20' and 11'20", to get the total reading of 286°31'20".

■ The Telescope

You take sightings using a telescope. The telescope is rigidly fixed to the traverse axis and revolves horizontally as well as vertically. An attached spirit level provides one way to take levels, but it's better to take levels with an engineer's level for the best accuracy.

Cross hairs in the telescope allow you to center on target, both horizontally and vertically. Two horizontal hairs, one above and one below the centered cross hair, allow you to take stadia measurement (see Chapter 6).

■ The Compass

The transit also has an attached compass, which allows you to give the direction of a bearing. As mentioned previously, beware of local attraction which can ruin your compass readings.

■ The Setup

The point where the transit is positioned, the *setup point*, is especially important to the survey. It's generally a lot corner or other point from which you can turn angles and lines sighted on course. You use the plumb bob suspended below the shifting head, the shifting head itself, and the tripod, to center the transit over the setup point.

First, press the tripod shoes firmly into the ground and bring the transit head as nearly level as your eye can judge. The plumb bob should be close to and barely clearing the tack in the stake on the ground marking the setup point.

Level the head and observe the position of the plumb bob with respect to the tack. It will seldom center the tack on the first try.

Loosen the level screws enough to shift the head toward centering the plumb bob and re-level. Repeat this process until the plumb bob is centered over the tack and the instrument is level.

■ Taking a Backsight

To use the telescope, setup the transit and take a sighting, called a *backsight*, to a reference point. Lock the two plates of the transit together on zero to zero.

After the telescope is sighted in on the reference point, clamp the transit head in place. Now you turn a 90° angle and measure a distance to locate a lost lot corner. Next, loosen the upper clamp and turn the upper plate to 90°, the angle required to find the desired corner (called a *foresight*), and then clamp the plate in place again. Use the tangent screw and magnifying glass to place the scales exactly on 90°.

Stretch the measuring tape from the tack under the plumb bob along the line of transit sight and set the required distance. Then set a stake and measure again. Finally, set a tack in the target stake at the exact distance and in the line of transit sight.

To *plunge* the telescope means to turn it 180° in a vertical circle. For instance, assume the transit is set up over a reference point as discussed above. This reference point is common to Lot 19 and Lot 20 of a typical subdivision. It is located at the northwest corner of Lot 19, which is the northeast corner of Lot 20. Lot 19 is staked, but the NW and SW stakes are missing from Lot 20. The back line of these lots is continuous.

From the reference point, backsight to the stake at the NE corner of Lot 19. Lock the transit head in place and plunge the telescope by turning it in a vertical circle to sight on the northwest corner of Lot 20. This sighting is exactly opposite (180°) the first sighting and gives the foresight that extends the back lot line of Lot 19, reestablishing the back lot line of Lot 20.

Measure 165 feet and set a stake with tack as previously explained. This is called *prolonging a line*. The back lot line of Lot 19 was prolonged to reestablish the back line of Lot 20.

Be sure to mark the direction that you turn an angle in your field notes. L20° means the angle was turned 20° to the left from the backsight target. Or, R20° means the angle was turned 20° right from the backsight target. Remember that a backsight isn't necessarily backward, behind you, or in a direction opposite the foresight. It is simply the sighting you take to a reference point before taking the second sighting.

■ Turning Successive Angles

If you need to turn a number of angles from the same reference point, turn them successively. Make a hand drawing in the field book to illustrate the angles you turn. Start by setting the vernier at zero. Then turn the first angle from the reference point (backsight target) to the desired point. Log this in the field book to correspond with the first angle in the illustration.

Turn the remaining angles successively and add each angle to the previous angle. That is, turn angle 2 and add it to the first angle. Turn angle 3 and add it to the first two.

Label each angle as it is turned and if each angle is to be turned from the reference point, set the vernier to zero each time and sight in the reference point. Also, if your transit has two verniers, do not use one then the other, stick to one for all readings.

Finally, before you turn the angle it helps to estimate it by eye. Is it less than 90°, or more than 90° and less than 180°? Is it greater than 180° but less than 270°? This will help you avoid reading the wrong number or the wrong vernier.

APPENDIX B

Geometry for Construction

Geometry is a branch of mathematics that deals with the measurement, properties, and relationships of points, lines, angles, surfaces, and solids. Thus, it is the basis of surveying. You can use geometric principles to lay out any shape and to find the measurements of all or any part of a natural or manmade figure. You need to know geometry for field work and mapping.

This section is a review of basic geometric tenets. If you need further brushing up on this math, study any beginning geometry or math review text.

The Line

The study of geometry begins with the straight line, which occurs when no definition of a curve is possible (although under certain conditions, a curve can be considered to be a straight line).

Straight lines form the sides of angles, triangles, rectangles, squares, pentagons (five sides), hexagons (six sides), and other figures that you measure and stake in construction layouts.

Draw straight lines using some form of straight edge, for instance, a drafter's triangle or a T-square. Establish two points and align the straight edge so that a line drawn along the edge will be drawn through both points (Figure B-1). Note the following three definitions.

1. *Parallel lines* never meet, no matter how far they are extended. Parallel lines are two straight lines drawn any distance apart (Figure B-2).

2. A *perpendicular line* is a line that intersects a second line in a way that forms 90° angles between the two lines (Figure B-3).

3. An *angle* is formed by two straight lines, called rays, that intersect in one point and one point only, and form an opening between the two rays (Figure B-4).

Angles

You measure angles using a protractor when you're in the drawing room (see Figure B-5) and by tape or transit when you're in the field (see Chapter 4). The following definitions apply to angles.

1. An *acute* angle measures less than 90° (Figure B-6 A).

2. An *obtuse* angle measures more than 90° but less than 180° (Figure B-6 B).

3. The measure of an angle depends only upon the size in degrees of the opening between the two rays (Figure B-6 C).

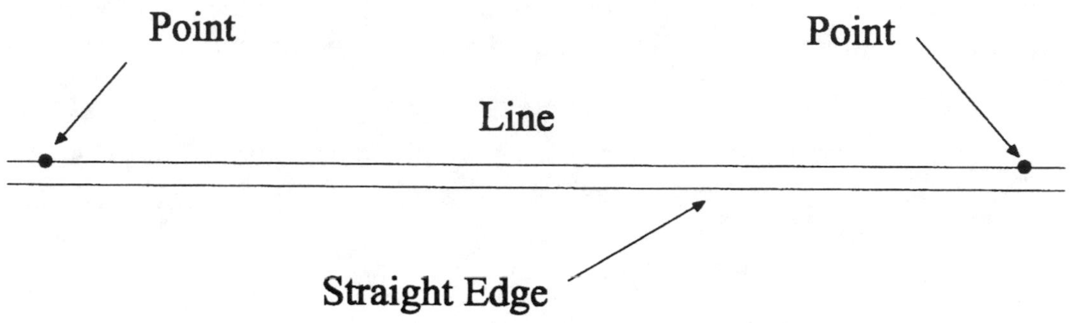

Point   **Point**

Line

Straight Edge

Figure B-1 *Drawing a line through two points*

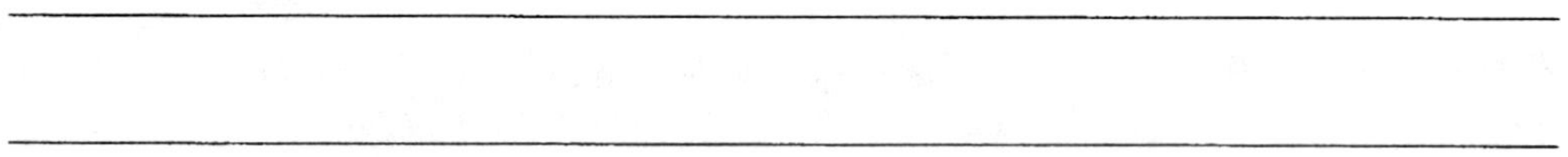

Figure B-2 *Parallel lines*

Triangles

1. A *triangle* is a plane figure bounded by three straight lines called *sides* (a *plane* is a flat or level surface). The sum of the three angles in every triangle equals 180° (Figure B-7).

2. The angles are identified by capital letters. The two acute angles are identified by *A* (the base angle) and *B* (the top angle). *C* identifies the third angle. *C* also identifies the 90° angle in a 90° triangle.

3. The sides are identified by the same letters in the lower case, with each side opposite its corresponding angle. That is, angle *A* and side *a* are opposite. Angle *B* and side *b* are opposite. Angle *C* and side *c*, the hypotenuse, are opposite. Sides *a* and *b* are known as the *legs* of the triangle.

4. Another important relationship between the angles and the sides is *opposite* side and *adjacent* side. These terms are used to identify the trigonometric ratios (see Appendix C).

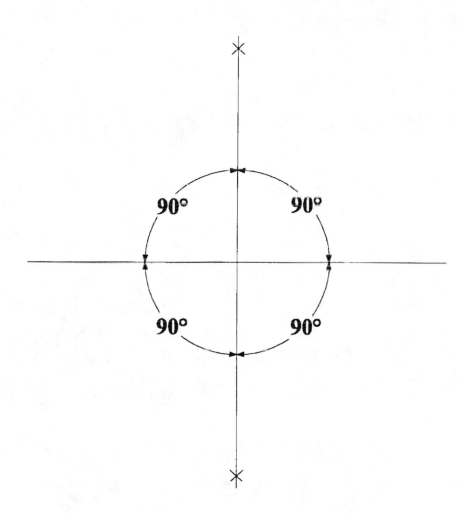

Figure B-3 *Perpendicular lines*

5. If you know two angles of a triangle, subtracting their sum from 180° gives you the value in degrees of the third angle (Figure B-8).

a. If you draw a line from the vertex, perpendicular to the base of a triangle, that is the *altitude* (Figure B-9).

b. The side upon which the triangle rests is the *base*.

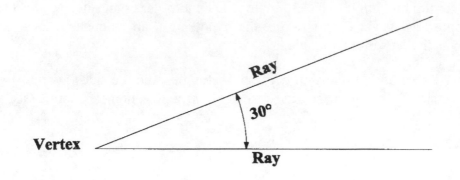

Figure B-4 An angle

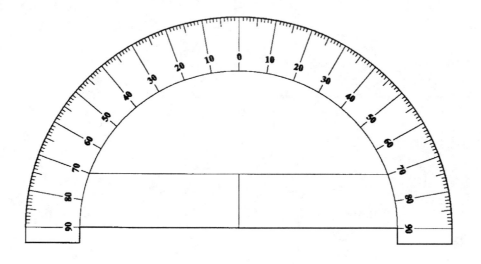

Figure B-5 A protractor

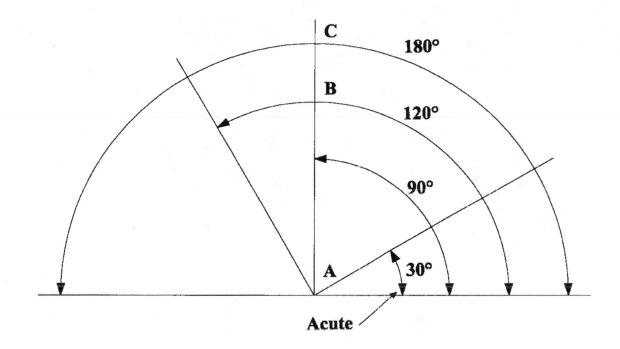

Figure B-6 *Various angles*

c. Either of the three sides of a triangle can be made the base. Each base of a right-angle triangle determines one of three possible altitudes (Figure B-10).

d. The type of triangle is determined by either the length of its sides or the degree of angle.

   i. The *right* triangle contains one angle of 90°. The size of the two remaining angles is determined by the lengths of the sides (Figure B-11). A relationship called the *Pythagorean theorem* determines the relationship between the three sides of a right triangle. The theorem states that the square on the hypotenuse equals the sum of the squares on the other two sides (Figure B-12). The *hypotenuse* is the longest side of the right triangle. Working relationships of right triangles demonstrate the use of square root calculations (Figure B-13).

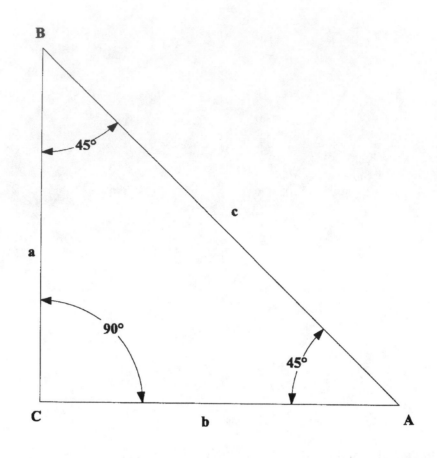

Figure B-7 *A triangle*

ii. An *isosceles* triangle has two equal sides. The angles opposite the equal sides are equal (Figure B-14). A line (altitude) drawn from the vertex, perpendicular to the third (unequal) side of the isosceles triangle, bisects the third side and forms two equal right triangles.

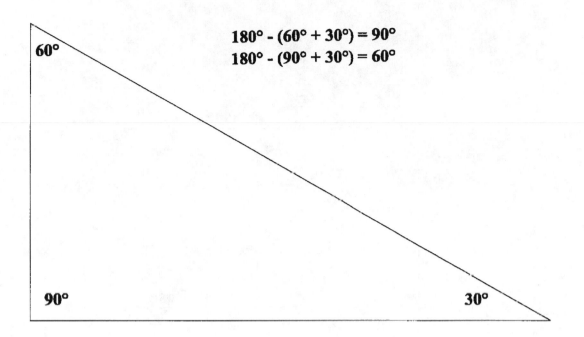

$$180° - (60° + 30°) = 90°$$
$$180° - (90° + 30°) = 60°$$

Figure B-8 *Finding the third angle*

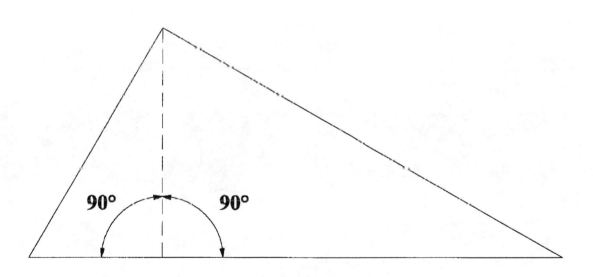

Figure B-9 *The altitude*

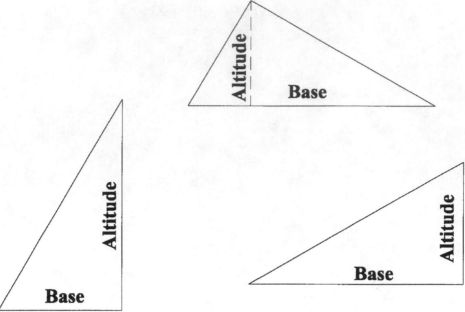

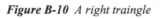

Figure B-10 *A right traingle*

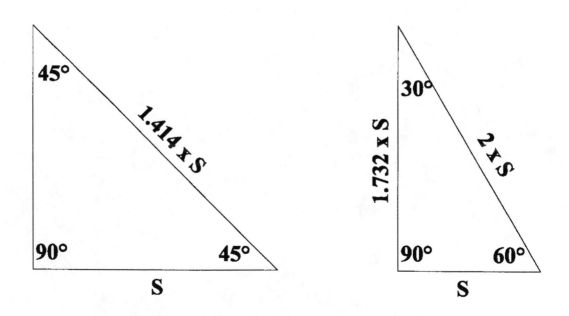

Figure B-11 *Defining the types of triangle*

iii. The *equilateral* triangle has three equal sides. Each angle equals 60° (180°/3 = 60°) (Figure B-15). The altitude drawn from any vertex of an equilateral triangle, perpendicular to the base, divides the triangle into two equal right triangles. The altitude is computed as: $a = 1/2s \sqrt{3}$. When the altitude is known, the sides can be computed as: $s = 2/3a \sqrt{3}$, where a = altitude.

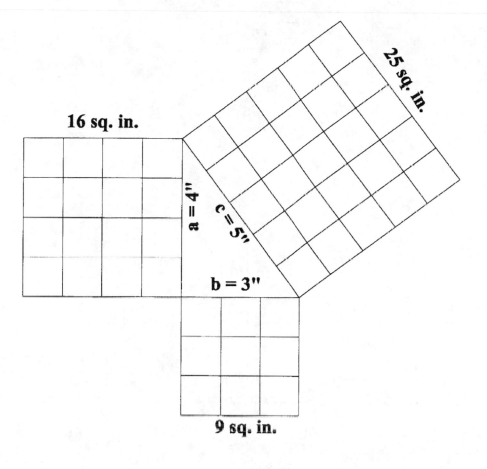

Figure B-12

$$c = \sqrt{a^2 + b^2}$$
$$c = \sqrt{16 + 9}$$
$$c = 5$$

$$a = \sqrt{5^2 + 3^2}$$
$$a = \sqrt{25 - 9}$$
$$a = 4$$

$$b = \sqrt{5^2 + 4^2}$$
$$b = \sqrt{25 + 16}$$
$$b = 3$$

Figure B-13 Square root calculations

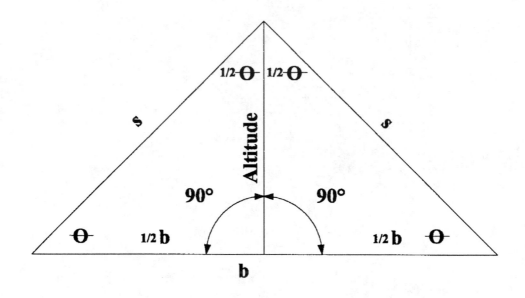

Figure B-14 Isosceles Triangle

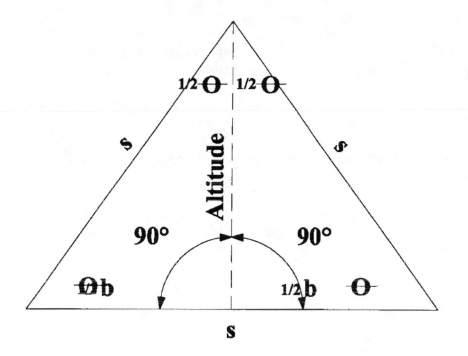

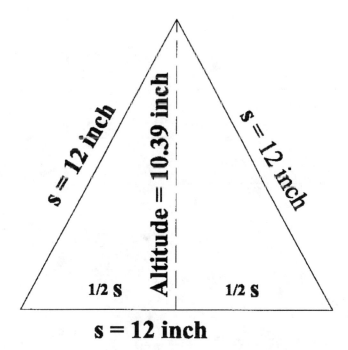

Figure B-15 Equilateral Triangle

iv. No two sides are equal in a *scalene* triangle and none of the angles of a scalene triangle equals 90° (Figure B-16).

e. The sides of a triangle generally are identified as shown in Figure B-17. Lower-case letters identify the sides, and capital letters identify the angles.

6. The area of a triangle is equal to one half the product of the base and altitude. That is, A = 1/2*ab* where *a* is the altitude and b is the base.

7. When you know only three sides of a triangle and you don't know the altitude and angles, you can find the area of the triangle by the formula

$$A = \sqrt{s(s-a)(s-b)(s-c)}$$

(see Figure B-18). The letter s in the formula represents one half the sum of the sides of the triangle. For instance, if

s = 1/2(a + b + c),

then,

s = (4 + 5 + 6)/2 = 7.5.

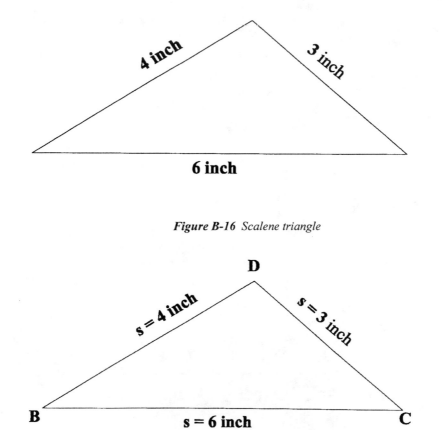

Figure B-16 *Scalene triangle*

Figure B-18 *Area of a triangle*

And,

$$A = \sqrt{7.5(7.5-4)(7.5-5)(7.5-6)}$$

$$\sqrt{7.5 \times 3.5 \times 2.5 \times 1.5} = 9.92 \text{ square inches.}$$

■ Useful Right-Triangle Relationships

There are three right triangles with relationships in each that are especially useful for layouts: the 30°-60°-90° triangle, the 45°-45°-90° triangle, and the 3-4-5 right triangle.

In the 30°-60°-90° triangle, the side opposite the 30° angle is one half the length of the hypotenuse. The side opposite the 60° angle is the square root of 3($\sqrt{3}$ = 1.732) times the side opposite the 30° angle.

In Figure B-19, the side opposite the 30° angle is equal to 1 unit. Then V × 1 = 1.732 × 1 = 1.732.

The 45°-45°-90° triangle is often called the *isosceles right triangle* since the two sides which include the 90° angle are equal (Figure B-20).

The 30°-60°-90° triangle and the 45°-45°-90° triangle are the drafter's triangles.

The 3-4-5 triangle has sides in the ratio of 3, 4, 5. Any numbers that are in this ratio will form a 3-4-5 triangle. Double 3-4-5 to get 6-8-10, or double 6-8-10 to get 12-16-20; the relationship is still true (Figure B-21).

For instance, you can readily find a square (90°) building corner or fence corner with these dimensions (in ratio) as in Figure B-14.

Quadrilaterals

Other geometric figures are in the category of *quadrilaterals*, which are plane figures bounded by four sides. The following definitions apply to quadrilaterals.

1. Remember that parallel lines lie in the same plane and never meet.

2. A *parallelogram* is a quadrilateral in which opposite sides are parallel, the opposite sides and opposite angles are equal, and the sum of the angles equals 360°.

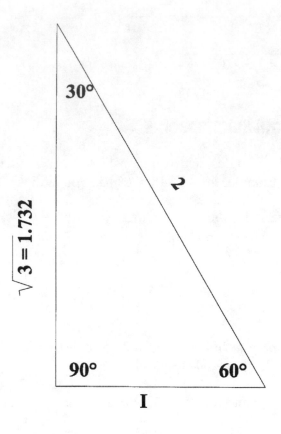

Figure B-19 *The 30-60-90 triangle*

3. If you draw a straight line joining any two opposite vertices, the line is called the *diagonal* of the parallelogram and it divides the figure into two equal triangles.

4. The perpendicular distance between the base and the opposite side is the altitude of a parallelogram.

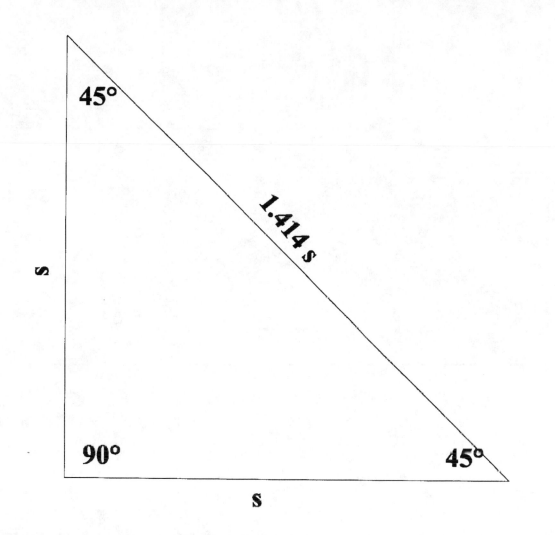

Figure B-20 *The 45-45-90 triangle*

5. A *square* is a parallelogram in which all sides are equal and all angles are right angles.

6. Two equal 45°-45°-90° right angles are formed by the diagonal of a square.

7. Since all sides are equal in a square, the area is equal to $s \times s = s^2$. $A = s^2$. For instance, in Figure B-22A let $s = 6$ inches.

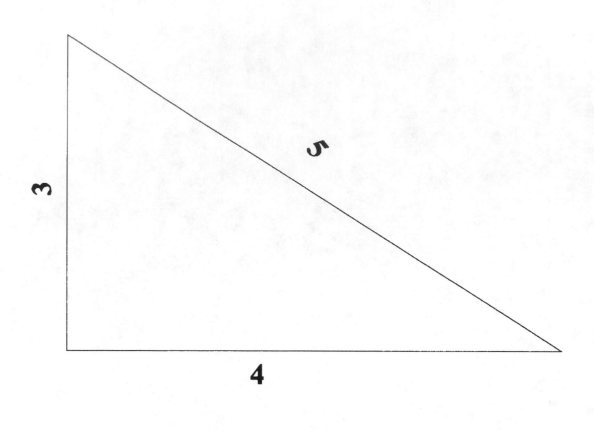

Figure B-21 *The 3-4-5 right triangle*

Then, the area of the square is $6^2 = 6 \times 6 = 36$ inches. The diagonal, which is the hypotenuse of a 45°-45°-90° triangle, is $d = s\sqrt{2} = 6\sqrt{2} = 6 \times 1.414 = 8.48$ inches. When you know the diagonal of a square, $s = 1/2d\ \sqrt{2} = 1/2 \times 8.48 \times 1.414 = 6$.

8. A parallelogram with four right angles (each 90°) is called a *rectangle* (Figure B-22B).

 a. The diagonal of a rectangle divides the rectangle into two equal triangles.

A.

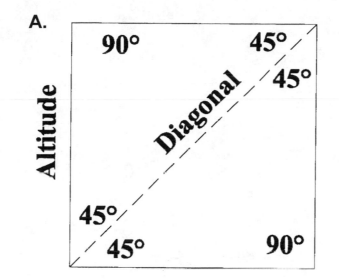

B.

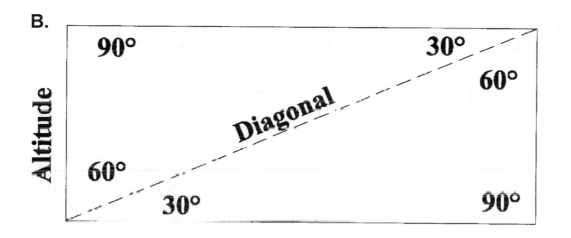

Figure B-?? *Area of a square and rectangle*

b. The area of a rectangle is equal to the altitude times the base: $A = ab$ where $a =$ the altitude and $b =$ the base.

c. The diagonal of a rectangle is a $\sqrt{a^2 + b^2}$ (Figure B-23).

d. Figure B-23 shows a parallelogram with sides parallel and the opposite angles equal.

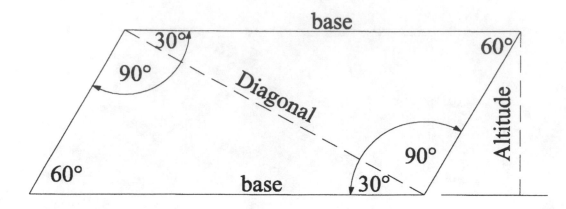

Figure B-23 *A parallelogram*

9. A quadrilateral with two sides parallel is called a *trapezoid*. Both parallel sides are called a base. The perpendicular distance between the bases is the altitude of a trapezoid (Figure B-24).

   a. The sum of the four angles equals 360°.

   b. In an *isosceles trapezoid* the two bases are parallel and the nonparallel sides are equal (Figure B-25).

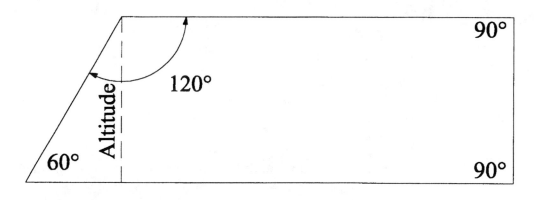

Figure B-24 *A quadrilateral*

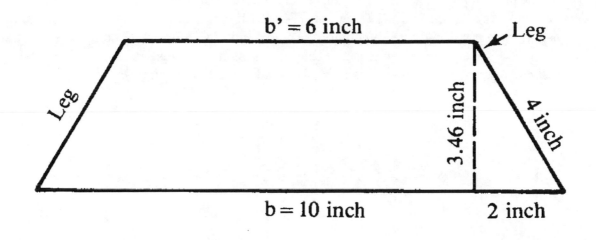

Figure B-25 *Isosceles trapezoid*

c. The area of a trapezoid is A = $1/2a(b + b')$. For instance, if you only know the bases given in Figure B-25, you need to find the altitude. Since the legs are equal, the center of b' lies on a perpendicular line from the center of b. This perpendicular line divides each base exactly into equal parts. In Figure B 25 b – 10 inches, b' = 6 inches, and a = 3.46 inches. Then, A – 1/2 $a(b + b')$ – 3.46(10 + 6)/2 = 27.68 square inches.

d. To find the altitude when you know only the bases: b = 10 inches and b' = 6 inches. The right half of b = 5 inches and the right half of b' = 3 inches. The base of the triangle formed by the altitude at the right end of the trapezoid is 5   3 – 2 inches. This is also true of the left end of the trapezoid.

e. The altitude of a trapezoid is the side of the triangle perpendicular to base b. If the leg of a trapezoid is the hypotenuse of the triangle (4 inches) and the base of the triangle is 2 inches, a = $\sqrt{4^2 - 2^2}$ = 3.464 inches according to the Pythagorean theorem.

Regular Polygons

A plane figure bounded by any number of equal sides and equal angles is called a *regular polygon*. For instance, an equilateral triangle is a regular polygon bounded by three sides, the square by four sides. Others, with sides 5 through 10, respectively, are the polygons called *pentagon, hexagon, heptagon, octagon, nonagon, and decagon* (Figure B-26).

1. You can indicate the number of sides by a numeral and the suffix *gon*. For instance, a pentagon is a 5-gon and a 17-gon is a 17-sided figure. For any number of *n* sides, an *n*-gon is an n-sided figure (a polygon). The polygon in Figure B-27 contains six equal sides and six equal angles. It is a regular polygon and because it has six sides, it is a *regular hexagon*. The diagonals form six equal equilateral triangles.

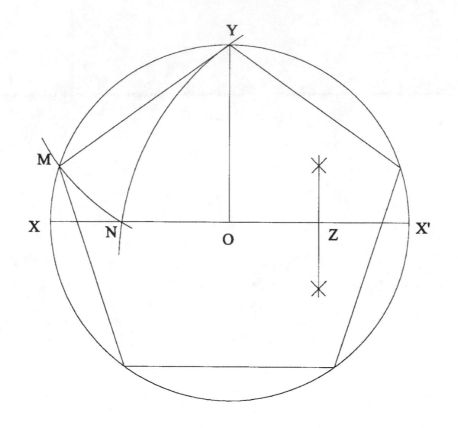

Figure B-26 *Regular polygons*

2. Since the altitude of an equilateral triangle is $a = 1/2s \sqrt{3}$, the altitude of the base triangle is a = $1/2 \times 6\sqrt{3} = 6/2 = \sqrt{3} \times 1.732 = 5.196$ inches.

3. To find the area, remember that there are six triangles in a regular polygon (as in Figure B-27). You know the area *A* of any triangle is $A = 1/2ab$. For the base triangle $A = 1/2 \times 6 \times 5.196 = 15.592$ square inches. So, $6 \times 15.59 = 93.532$ square inches, the area *A* of the regular polygon.

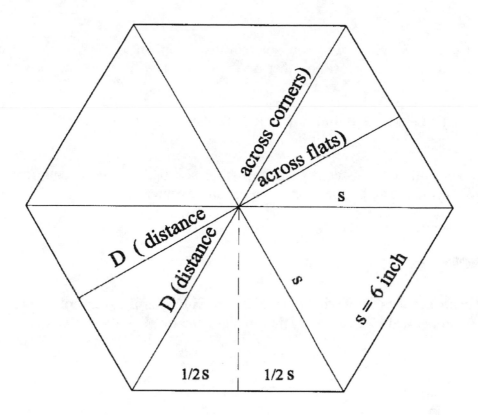

Figure B-27 *Triangles of a polygon*

4. Besides the length of side and the degree of angle in the hexagon, two other measurements are important: The distance *across corners* and the distance *across flats*. For instance, in Figure B-27, D1 is the distance across corners and D2 is the distance across flats. To lay out a hexagon, you can use either of these two distances. For instance, in Figure B-28, using distance *AB* across corners as a diameter, draw a circle. With the radius of the circle and points *A* and *B* as centers, draw two arcs, each intersecting the circle in two points. Connect all points around the circle, as shown.

Using the distance across flats as a diameter, draw a circle. Draw the vertical diameter. Draw a line perpendicular to the vertical diameter at the diameter end. Repeat at the other end. The sides of the hexagon are at 60° as shown, and tangent to the circle.

Tangents

A *tangent* is a line, either straight or curved, that is drawn to a curve so that it passes through two points infinitely close together on the curve. The line touches the circle at the *tangent points*.

1. If the tangent is a straight line, it is perpendicular to the circle's diameter at the tangent point.

2. If the tangent point is established by two circles, it lies on the circumference of both circles and on a straight line connecting the circle centers.

Conic Sections

The four *conic* sections shown in Figure B-29, the circle, ellipse, parabola, and hyperbola, are *plane curves* obtained by cutting a right-circular cone by planes at different angles to the cone.

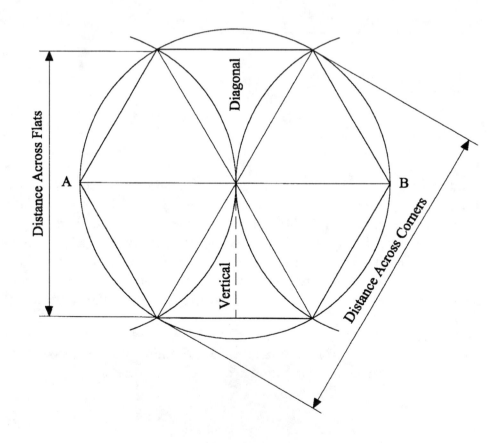

Figure B-28 *A hexagon*

Prisms

A *prism* is a solid that has parallel polygons for bases and faces (sides and ends) that are parallelograms.

1. A prism with rectangles as bases is called a *rectangular prism* (Figure B-30).

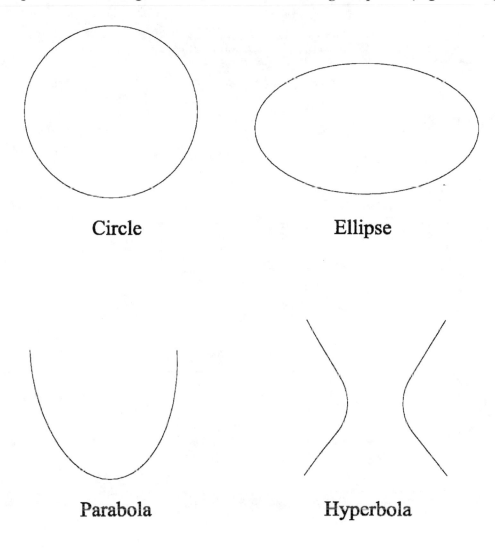

Circle   Ellipse

Parabola   Hyperbola

Figure B-29   *Four conic sections*

2. A *right triangular prism* is one that has edges that are perpendicular to the bases, and the bases are triangles (Figure B-31).

3. A *cube* is a rectangular prism in which all six faces are equal squares. Various forms of prisms enter into cut-and-fill computations for earth moving.

4. The volume of a right prism is equal to the area of the base multiplied by the altitude. If the prism is a cube, any side can be a baseline or the altitude (Figure B-32). For example, if you want to find the volume of a cubic form with sides s equal to 4 feet: V $= s^3$. $V = 4^3 = 64$ cubic feet. Or, let V equal the volume of a prism, A equal the area of the base, and a equal the altitude. Then, $V = A \times a$.

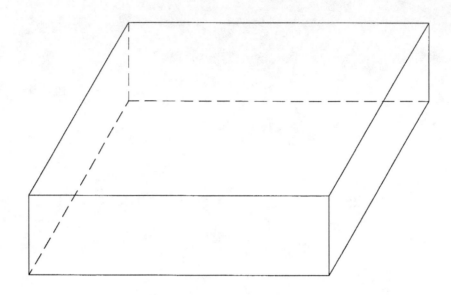

Figure B-30 *A prism*

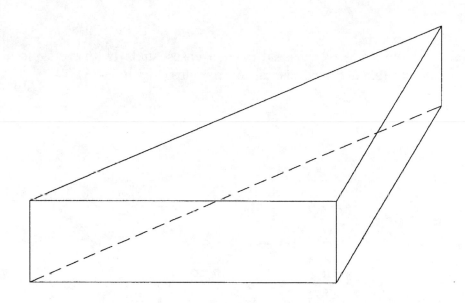

Figure B-31 *A right triangular prism*

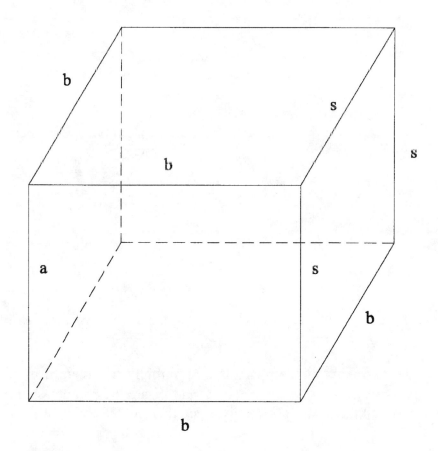

Figure B-32 *A cube*

Prismoids

A *prismoid* is a solid with two unequal parallel bases (ends) with the same number of sides that has other faces that are quadrilaterals or triangles (Figure B-33).

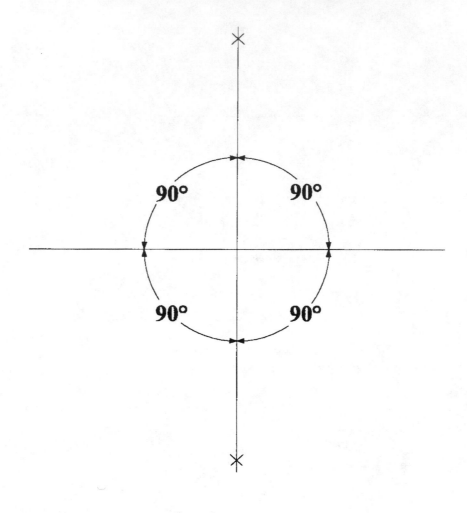

Figure B-33 *A prismoid*

1. A simple formula commonly used for finding the volume of a prismoid is known as the *average-end method.*

   Average the two ends of the prismoid and multiply the average by the length. The formula is V = $[(A^1 + A^2)/2] \times$ length. A^1 and A^2 are the end areas and length l is the distance between them. $(A^1 + A^2)/2$ gives you the average end area.

2. A formula that gives the exact volume of a prismoid is $V = (A^1 + 4Am + A^2)/6 \times \text{length}$. The new term, Am, is the area of the section at middle distance from the ends. Find Am by averaging the length and width of the ends.

3. Add the lengths of the bottom and top and divide by 2 to get an average midpoint dimension.

Circles

The area of a circle is $A = \pi r^2$. $\pi = 3.1416$ and $r = \text{radius}$.

The ratio of the areas of two circles is the same as the ratio of the squares of their radii, the squares of their diameters, or the squares of their circumferences.

Practice Problems

1. Find the area of the right triangle in Figure B-34. From Figure B-12, the square root of $10^2 - 5^2$ gives you the third side, which is the altitude of this triangle.

$$a = \sqrt{10^2 - 5^2} = \sqrt{100 - 25} = \sqrt{75}$$

$$= \sqrt{25 \times 3} = 5\sqrt{3} = 5 \times 1.732 - 8.66 \text{ inches.}$$

Since the altitude is 8.66 inches, the base is 5 inches, and the area is $A = 1/2ab$, then $A = 1/2 \times 8.66 \times 5 = 21.65$ square inches. Sometimes this is written as $21.65''^2$.

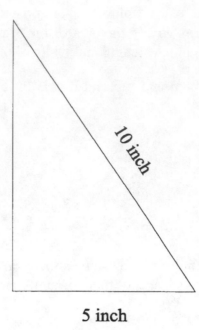

10 inch

5 inch

Figure B-34 *Find the area*

2. Inscribe (draw one figure within another) an equilateral triangle in a circle (Figure B-35). The straight line distances *AB*, *BC*, *CD*, *DE*, and *EF* are chord lengths, each equal to the radius of the circle. Draw the circle.

   Measure off the chord lengths. Connect alternate points *AC*, *CE*, and *EA* as shown. This is the equilateral triangle.

3. Inscribe a circle inside an equilateral triangle. The bisectors of the angles of a triangle are concurrent (intersect) in a point equidistant from the sides of the triangle. The bisectors of the sides of a triangle are concurrent in a point equidistant from the sides of the triangle.

   Draw the equilateral triangle to any size (Figure B-36). Drop a line *CD* from the vertex *C* and perpendicular to base *AB*. Line *CD* bisects the base *AB*.

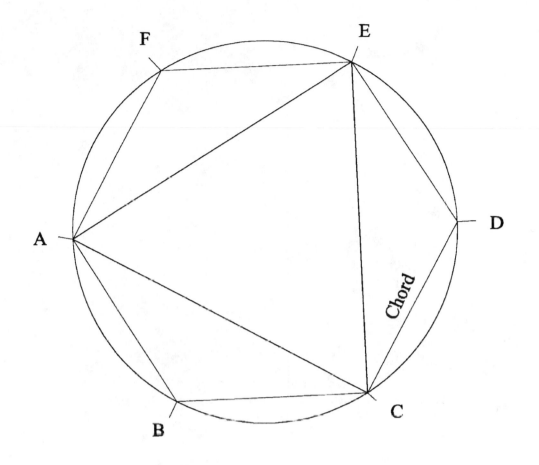

Figure B-35 *Inscribe a triangle*

Bisect side *AC* as shown. This bisecting line intersects line *CD* in the center of the triangle. Draw the circle with radius *OE*. The sides of the triangle are tangent to the inscribed circle.

4. Find the volume of a right prism with a rectangular base of 10×5 inches and an altitude of 7 inches.

   $V = A \times a = 10 \times 5 \times 7 = 350$ cubic inches. Since 1 cubic foot equals 144 cubic inches, $350/144 = 2.43$ cubic feet.

5. Find the volume of a right triangular prism with bases 3×9 feet and an altitude of 5 feet.

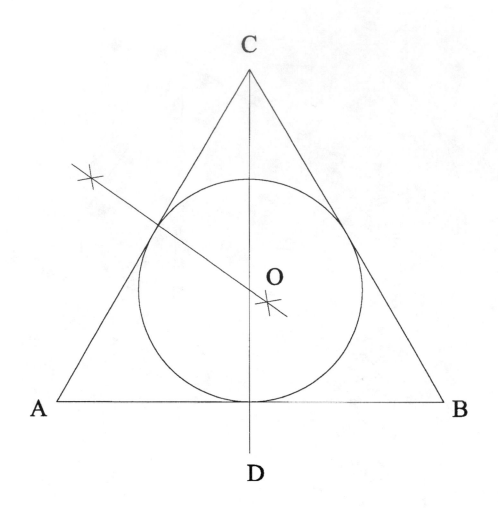

Figure B-36 *Replacing pipes*

The area of a triangle is A $= 1/2ab$. In this case A $(9 \times 3)/2 = 13.5$ square feet.

Then, $V = A \times a = 13.5 \times 5 = 67.5$ cubic feet.

6. You need to replace a 30-inch pipe with four smaller pipes.

What size should the smaller pipes be? $D^2 = 30^2 = 900$. Since four pipes are required to replace one pipe with a diameter squared of 900, then $900/4 = 225$, the diameter squared of one of the replacement pipes. $\sqrt{225} = 15$, the diameter of each replacement pipe.

7. Find the altitude of the equilateral triangle in Figure B-15B.

$a = 1/2s \sqrt{3}; a = 1/2 \times 12 \times 1.732 = 10.39.$

When you know the altitude, you can find the side of an equilateral triangle by transposing the formula.

$$a = 1/2s\ 3;$$

$$a/(1/2) = s\sqrt{3}; a \times 2/1 = s\sqrt{3};$$

$$2a = s\sqrt{3}; s = 2a\ /\sqrt{3};$$

$$s - 2a/\sqrt{3} \times \sqrt{3}\ /\ \sqrt{3};$$

$$s = 2a\ \sqrt{3}/3; s = 2/3a\ \sqrt{3}.$$

8. Find the side of an equilateral triangle with an altitude of 15.59 inches.

$$s = 2/3a\ \sqrt{3}.$$

$$s = 2/3 \times a \times \sqrt{3}.$$

$$s = 2/3 \times 15.59 \times 1.732 = 18\ \text{inches}.$$

APPENDIX C

Trigonometry for Construction

Trigonometry is the study of the properties of triangles and of the trigonometric functions of their angles. In surveying, you must be able to find the size, area, and angles of a triangle. This information will come in handy in every survey you do.

Trigonometric functions are ratios used to solve for parts of a triangle. The trigonometric functions used in surveying are called sine, cosine, tangent, secant, cosecant, and cotangent. Respectively, they are abbreviated sin, cos, tan, sec, csc, and cot.

For example, in Figure C-1, side a is opposite angle A and adjacent to angle B. Side b is opposite angle B and adjacent to angle A. The sine of the acute angle A equals the ratio of the side opposite (side a) this angle to the hypotenuse.

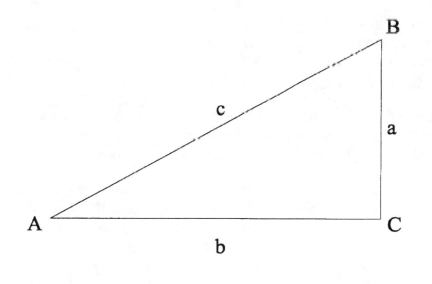

Figure C-1 *Trigonometric functions*

The cosine of the acute angle A equals the ratio of the side adjacent to this angle to the hypotenuse. In the same way, the sine and cosine relate to angle B; side b is its opposite and side a is adjacent.

The six trigonometric functions are based on this relationship between the sides and the acute angles as follows.

$$\sin A = \frac{\text{side opposite}}{\text{hypotenuse}} = \frac{a}{c} \qquad\qquad \sin B = \frac{\text{side opposite}}{\text{hypotenuse}} = \frac{b}{c}$$

$$\cos A = \frac{\text{side adjacent}}{\text{hypotenuse}} = \frac{b}{c} \qquad\qquad \cos B = \frac{\text{side adjacent}}{\text{hypotenuse}} = \frac{a}{c}$$

$$\tan A = \frac{\text{side opposite}}{\text{side adjacent}} = \frac{a}{b} \qquad\qquad \tan B = \frac{\text{side opposite}}{\text{side adjacent}} = \frac{b}{a}$$

$$\sec A = \frac{\text{hypotenuse}}{\text{side adjacent}} = \frac{c}{b} \qquad\qquad \sec B = \frac{\text{hypotenuse}}{\text{side adjacent}} = \frac{c}{a}$$

$$\csc A = \frac{\text{hypotenuse}}{\text{side opposite}} = \frac{c}{a} \qquad\qquad \csc B = \frac{\text{hypotenuse}}{\text{side opposite}} = \frac{c}{b}$$

$$\cot A = \frac{\text{side adjacent}}{\text{side opposite}} = \frac{a}{c} \qquad\qquad \cot B = \frac{\text{side adjacent}}{\text{side opposite}} = \frac{a}{b}$$

You can use a table known as the table of trigonometric functions to find the six trigonometric functions for all acute angles between 0° and 90°.

These ratios are abstract numbers that can be expressed to varying numbers of decimal places for varying degrees of accuracy. For most purposes, you'll find numbers to five decimals adequate. For very accurate work, you could use tables of eight or more decimal places.

For example, in a four-place table sin 18°10' is .3118. In an eight-place table sin 18°10' is .31178219. In the four-place table the fourth digit rounds off one greater than the fourth-place digit in the eight-digit table due to the value of the fifth digit in the eight-place table.

The small circle at the top right of a number (19°) indicates *degrees* in an angle. The prime mark at the top right of a number (10') indicates *minutes* of a degree. *Seconds* (of a minute) are indicated by a double-prime mark at the top right of a number (30"). Sixty seconds equal 1 minute of a degree. Sixty minutes equals 1 degree.

Practice Problems

You can find tables that list the functions for degrees, minutes, and seconds. The following examples illustrate how to use such a table.

1. Find the values of the six trigonometric functions for angle 10°20'. Since this angle is less than 45°, find 10°20' in the left-hand column. The values horizontally across the table are:

	sin	cos	tan	cot	sec	csc	
10°20'	.17937	.98378	.18233	5.4845	1.0165	5.5749	80°40'
	cos	sin	cot	tan	csc	sec	

2. Find the values of the six trigonometric functions for angle (or bearing) 80°40'. This is the complementary angle to angle 10°20' since the two angles total 90°.

    Find 80°40' in the right-hand column. The headings are at the bottom of the page as shown above. Note, for example, that sin 10°20' equals cos 80°40'. When the value of the acute angle is greater than 45°, use the right-hand column and read from the page's bottom upward. If the tabular value is given and the angle corresponding to that value is given, find that value in the table and read the angle that you want.

3. Find the value of an angle with sin .17937. First, find .17937 in the column headed sin at the top of the table. In the left-hand column read 10°20' for the value of the angle.

4. Find the value of an angle with sin .98378. You'll find the value .98378 in the column headed sin at the bottom of the table. In the right-hand column you'll read 80°40', the value of the angle.

5. If a table lists functions in increments of, say, 10 minutes, you'll need to interpolate. For instance, find the sin of angle 10°25' with a table that lists degrees only in increments of 10' or, 10°20', 10°30', 10°40', and so on. You need to interpolate within the 10' spread of values. The sin increases in value from 10°20' to 10°30'. For each 10' the sin increases in value. Then,

    Sin 10°30' = .18224
    Sin 10°20' = .17937
    Difference = .00287

    The difference, .00287, represents a tabular value of 10'. Since the value you want is for 25', which is halfway between 20' and 30', add one half of .00287 = .001435 to sin 10°20'. Or,

    17937 + .001435 = .180805.

    Sin 10025' is .180805.

Another way to interpolate this type of problem is to consider only 287 from the difference. That is, if 287 represents 10' of a degree, then one tenth of 287, or 28.7, represents 01' (1 minute) of a degree.

For 5' of a degree, the value is 5 × 28.7 = 143.5. Place the two zeros dropped from .00287 before 143.5 for .001435. This is the value for 5' of a degree as required by interpolation. Then .17937 + .001435 = .180805, the sin of 10°25' previously found.

6. Suppose you need to lay out 12 equally spaced points on a 60-foot diameter circle in order to place columns for a rotunda. You know that if the diameter is 60 feet, the radius is 30 feet.

Since you need 12 equally spaced points around the circle, the angular distance measured on the arc between adjacent points is 360/12 = 30°.

Draw radii *OA* and *OB* from the center and connect them to the chord of the arc *AB* to form an isosceles triangle. The bisector of the 30° central angle is perpendicular to and bisects the chord, forming the two equal triangles *AOC* and *BOC*.

Use trigonometric methods to find either *AC* or *BC* (each is one half the length of the chord) and lay out the points. Two times either dimension is the length of the chord.

Again, to find the length of the chord, multiply the diameter by the sine of one half the central angle: 60 × sin 15°.

When you draw the perpendicular *OC*, the 30° central angle is divided into two 15° angles. The triangle of Figure C-2 is formed. Angle *AOC* = 15° and side *AO* = 30 feet. Then, sin *O* = *AC*/30 = .25882; *AC* = 30 × .25882 = 7.76 feet to two decimal places, or 7.7646 feet to four decimal places.

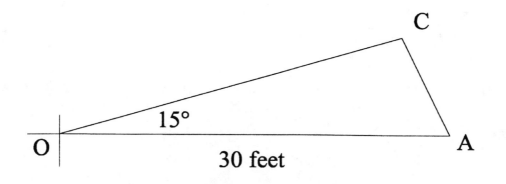

Figure C-2 *The triangle is formed*

The chord *AB* is $2 \times 7.7646 = 15.5292$ feet, about 15.53 feet. To find the length of the chord using the diameter, multiply the diameter by $\sin 30/2 = 60 \times \sin 15° = 60 \times .25882 = 15.5292$ feet.

APPENDIX D

More Practical Examples

Using what you've reviewed about using a transit and how to apply the principles of geometry and trigonometry to surveying, try the following examples for further problem-solving practice.

1. Suppose you need to measure on a slope like the one shown in Figure D-1. You'll need to find the horizontal plane *AB* first. How?

    Start by establishing a plane that's vertical to the earth's surface. This is simple: just suspend a plumb bob. The plumb bob will point directly toward the earth's center establishing a vertical line (*BB'*) with respect to the earth's surface. Now, if you establish a plane that's perpendicular to the vertical line (*BB'*), you'll have the horizontal plane *AB*.

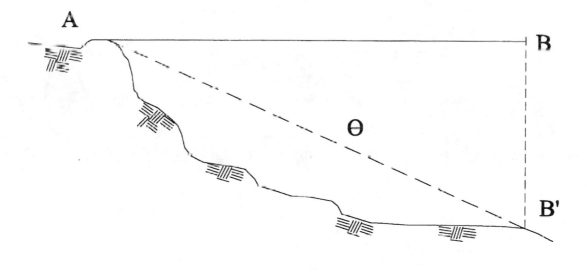

Figure D-l Measuring on a slope

Measure the triangle you've established. If the distance *BB'* is not too great, you can take the measurement with a tape. Otherwise, take it as a series of measurements so that the vertical distance at the leading end of the tape is never too great to handle. Or, measure *AB'*, measure angle e, and compute the distance by trigonometric methods.

2. If the area is hilly, the vertical distance may be so high that you'll need to take several measurements to complete the total distance. Assume your site is the one shown in Figure D-2. You need to take a series of measurements down a slope (it is much easier to measure down-slope than up-slope).

    For precise measuring, place a level on the tape and use a spring balance showing pounds of pull to maintain uniform tension in the tape when you extend it to measure.

    You'll get a precise measurement between points *AB'* if the intervening points are well aligned. The alignment is best done by sighting with a transit. Figure D-2 shows that careless alignment will give you total measurement along the solid line that is greater than the actual distance.

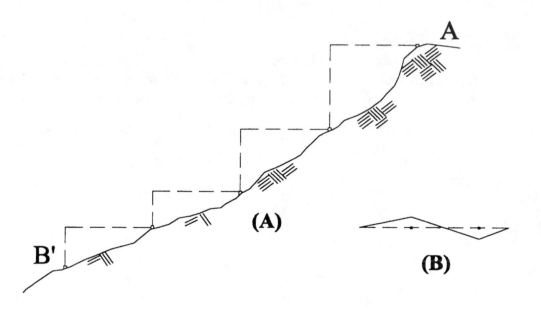

Figure D-2 *Measuring in a hilly area*

Sometimes you can't see or reach the beginning and end points of the measurement. That's when you'll need to use indirect measurement and computations by geometric proportion. For instance, in Figure D-3 you need to find the distance between points *A* and *B* on opposite sides of a knoll.

First, establish point 0 at any convenient place. Make $OA' = OA$ and $OB' = OB$ such that $AB = A'B'$. Now you can solve for the triangles.

4. What if you can set only one point? Try using the method shown in Figure D-4.

Set point A at any convenient point with $AB = AC$. Then, establish points D and E at any convenient distance with $AD = AE$. And simply calculate BC as $BC = (DE \times BA)/AE$ (BC equals the product of DE and BA divided by AE).

This method (or the one in problem 3) would work well if you had to take measurements across a pond.

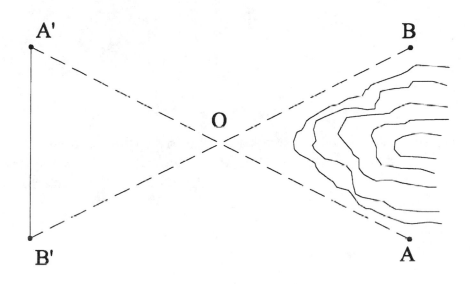

Figure D-3 *Opposite points on a knoll*

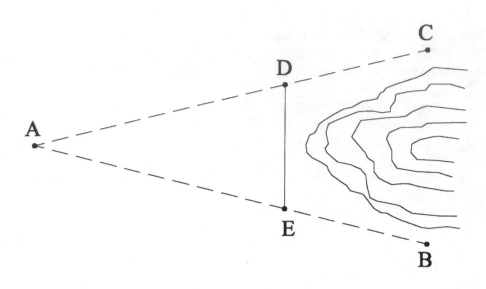

Figure D-4 *Opposite points-alternative method*

5. Suppose you need to measure across a creek or river without crossing the stream. Use the method shown in Figure D-5, which uses a tape measure to establish the 60° angle. Then, using the principles relative to the 30°-60°-90° triangle, measure the distance *DP*.

   Use any convenient distance, for instance 130 feet, for *DE* and *CF* to lay out *EF* parallel to *DC*. Then, *CB* is parallel to *DP*.

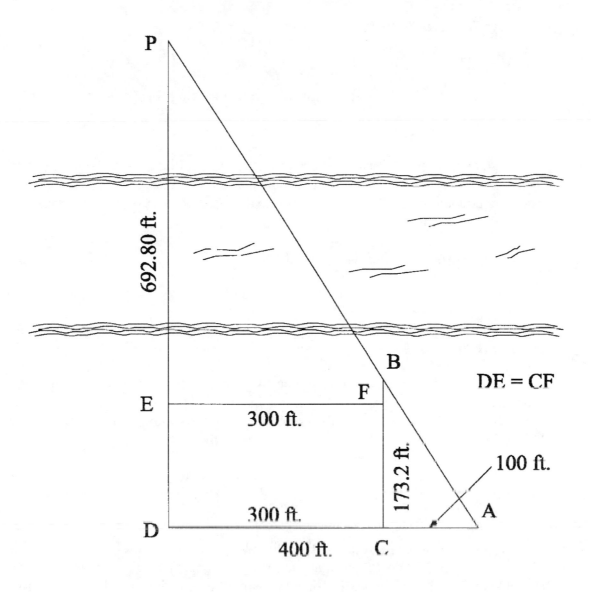

Figure D-5 *Measuring across a stream*

Next, establish *AC* at 100 feet and *CB* at 173.2 feet. *AB* measures 200 feet. (Remember the relationship of the sides of a 30°-60°-90° triangle).

Finally, angle *CAB* = 60° and triangle *ADP* has the same relative proportion as triangle *ACB*. So,

AD = 400 feet.

$DP = 1.732 \times 400 = 692.8$ feet, and

$AP = 800$ feet.

So, side $DP = 692.80$ feet, the distance you're looking for.

6. Suppose you need to find the height of point B on a building (Figure D-6).

First, assume point A is at the center of the transit. Place a level sight on point C and use a level rod to measure the distance 5 feet, 6 inches to the street. Second, turn angle O and find 32°30'. Then measure distance AC to get 240 feet. Using the principle of trigonometry,

$\tan \theta = BC/AC$; $\tan 32°30' = BC/240$; and

$.63707 = BC/240$; $BC = 240 \times .63707 = 152.89$ feet.

If you add, $152.89 + 5.5 = 158.39$ feet, giving you the total distance from street level to point B.

Sometimes you can use only a tape to get the measurements you need. For instance, suppose you are buying the lot in Figure D-7. After you examine the deed record to find the size of Lot 29, use a tape survey to verify the lot dimensions and make a few calculations to verify the acreage.

First, in Lot 29, choose point P as a reference point. Measure and record the distance from each corner of the lot to point P. Then measure and record the lot lines.

As shown, you've established five triangles. The sum of the five areas equals the area of Lot 29. You'll see that the triangle adjacent to Tudo Street has two equal sides. By the principle of geometry, with 60 feet as the triangle base, the two base angles each equal 45°.

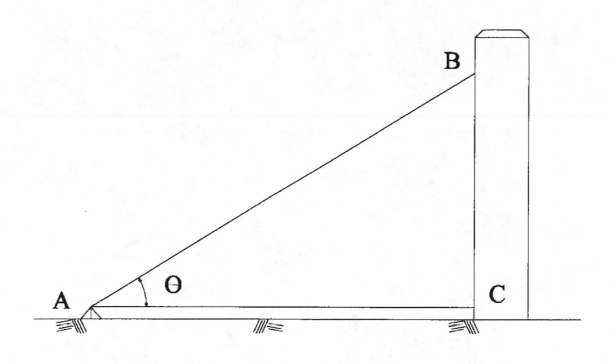

Figure D-6 *Finding the height of a point on a building*

Then draw a perpendicular line from the vertex *P*, to the base, *AE*. This bisects the base and forms two equal triangles. The hypotenuse of each triangle is then 42.36 feet. You know that when a 45° triangle is involved, the altitude times 1.414 equals the hypotenuse or,

hyp = a × 1.414.

Then,

a = hyp/1.414 = 42.36/1.414 = 29.96 feet.

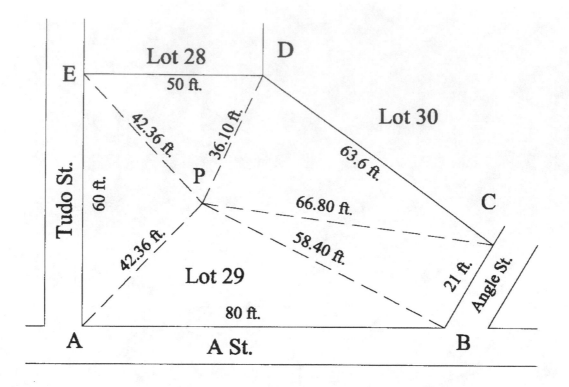

Figure D-7 *The lot*

The area of a triangle is altitude times base divided by 2. So,

$$A = 0.5ab = 29.96 \times 30/2 = 449.4 \text{ square feet.}$$

Since there are two equal triangles in the area enclosed by triangle *APE*, the area of triangle *APE* is $2 \times 449.4 = 898.8$ square feet. Next, use the formula $A = \sqrt{s(s-a)(s-b)(s-c)}$ to solve the remaining triangles.

For triangle *APB*,

$$s = 0.5 (80 + 42.38 + 58.40) = 90.39 \text{ feet.}$$

Then,

$$A = \sqrt{\frac{90.39 (90.39 - 80)(90.39 - 42.36)}{(90.39 - 58.40)}}$$

$$= \sqrt{90.39 \times 10.39 \times 48.03 \times 31.99}$$

$$= \sqrt{1{,}442{,}988.13}$$

$$= 1201.24 \text{ square feet, the area of } APB.$$

Finally, solve the remaining three triangles and add all the areas to get the total area of lot 29 in square feet. Then, divide the number of square feet by 43,560, the number of square feet in an acre, and you'll have the area of lot 29 in acreage.

7. Locating objects by offsets (measurements taken from a given base point or base line) is another easy and quick method you can do using a tape. For instance, Figure D-8 shows a building and a small stream. The base reference point is either in the center line or the east right-of-way line of County Road 2. Here it is in the east right-of-way line. Set the first offset 198 feet from this point.

The dimensions are shown between offsets along the north right-of-way line of State Road 764. These offsets are the lines at 90° angles. The length of each offset is given to the building or stream.

Use 1-inch iron-pipe stakes 3 or 4 feet long to mark reference points along the stream for future use. If you get the offset lines at 90° to the reference line, you can establish the 90° angle by the 3-4-5 triangle method. Try using 30, 40, and 50 feet for the triangle sides.

Establish the building corners. Points along the stream should, in general, fall near or at a turn in the stream (this way, the meandering of the stream is mapped). Or, you could establish random points such as *A, B, C, D, E,* and *F,* and take two measurements from them to one building corner or stream marker.

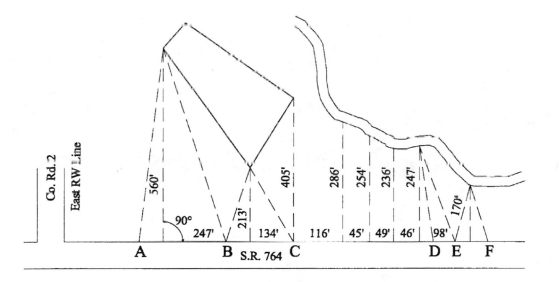

Figure D-8 *Offsets*

In this way, you can set the stream markers beforehand and avoid the tedious work of turning the 90° angles to the reference points (see the non-dimensioned lines in Figure D-8). These are called *oblique measurements*.

Note: usually you'll use *hubs* (2-inch square stakes) driven almost to ground level and set a surveyor's tack in the hub at the point of reference. Sometimes you'll leave the hubs as markers. Or, you can pull them out and use a more durable iron or concrete marker.

Many county or regional planning commissions establish the type and size of marker you should use for a specific purpose. If you use plastic, make sure it's capped with metal so it can be located with a metal detector.

INDEX

A

Abscissa.. 165
Acute angle................................ 194, 195, 225
Apex of angle 187

B

Building sites..................................... 20, 118

C

Calibrating the stadia wire............................ 93
Characteristics of an ellipse 75
Choosing a site 3, 20
Chords and angles................................... 130
Circles... 219
Complementary angle.................................. 71
Conic sections...................................... 214
Considerations in the field.......................... 30
Contour intervals.................................... 97
Correction factors................................ 86, 87
Cross-levels 112
Curve 130

D

Deed 14
Defining curves in the field 45
Deflection angle 48, 50, 136
Differential leveling................................ 112
Direct-contour method................................ 99
Dividing a line into equal parts 3, 68
    Radius 80
Double areas.................................... 29, 34
    Meridian distance 33
Drainage 23
Drawing a tangent circle........................ 3, 69

E

Easements... 23
Economizing in the field 3, 34
Ellipses... 4, 75
Equal altitude method 178
Equal-tangent curve.............................. 140

E (continued)

Equilateral triangle 201, 211, 212, 220, 222
Establishing Parallel Lines 3, 67
Evaluating a Site 4, 118

F

Field work........25, 68, 99, 129, 145, 167, 168, 193
Finding departures 3, 4, 31, 34
    Elevations................................. 4, 110
    Latitudes..................................... 31
    Missing measurements......................... 148
Flying levels 112

G

General Points 4, 169
Geometry for Construction.............. 5, 185, 193
Grade Stakes 4, 113
Grading 4, 123

H

Heating... 22
Hexagon... 4, 74
Horizontal angle 178, 179
    Sighting 90

I

Illustrating a depression 4, 96
Initial points 10
Intersection angle.............................. 130
    Of lines 44
Involute of a circle 4, 78
    Of the pentagon.............................. 78
Isosceles triangle 198

K

Keeping notes 112

L

Land deeds 3, 14, 27
    Divisions 10
    Surveying 9
Laying out a line.............................. 3, 51
    Out angles................................. 3, 71

Out curves 133
Out roads 3, 51
Length of curve 132
Level grade 113
Leveling instruments 105
Locating by latitude and longitude 5
Locating true north 5, 178

M

Making a topographic survey 4, 97
Mapping procedure 4, 168
Meridians 10, 11
Meridional boundaries 11

N

Necessary curves 4, 128

O

Obtuse angle 194
Original surveying notes 145

P

Parabolic curve 139, 141
Parallel lines 193, 194
Parallelogram 205
Parallels of latitude 10
Pentagons 4, 77
Perpendicular line 3, 65-67, 193, 237
Planning a bridge 61
    The curb 120
    The sewers 120
    The streets 4, 118
Plotting angles 168
Potential for error 4, 115
Practice survey 3, 27
Preliminary survey 21
Prismoids 218
Prisms 215
Pythagorean theorem 197, 211

Q

Quadrilaterals 205

R

Radius 130
Range lines 11
    Of townships 11
Reading levels 106
Reciprocal leveling 113
Recording the measurement 92
Rectangle 31, 50, 51, 163, 165, 166, 182, 208, 209
Rectangular system 10
Reference lines 3, 44, 97
Regular polygons 211
Reticules 84
Right angle 207
Road repair 23
Rods 83
Running the survey lines 27

S

Scalene triangle 204
Schools 23
Setting parallel lines 3, 57
    Stakes 133
    Up the level 105
Sewerage systems 22
Sighting along an incline 91
Similar triangles 85
Simple curves 128
Spiral of archimedes 79
Stadia 4, 26, 83
    Method 83
Straight lines 193
Street and alley intersections 138
Subchords 134
Subdividing land 4
Survey precision 40
Surveying instruments 3, 26
Surveyor's level 26

T

Taking sightings 178
Tangent distance 131
Tangents 172, 214
Tape measure 26, 234

Townships 11, 12
Transit 3, 5, 26, 57, 185, 187
Triangles 194
Trigonometry for construction 5, 185, 225
Turning successive angles 191
Type of development 3, 20
Types of leveling 4, 112

U

Unequal-tangent curve 142
Using a protractor 5, 172

An established point 3, 66
Hand Levels 107
Trigonometry 5, 172
Utilities 22

V

Vertical curve 139, 143, 144

W

Waste disposal... 23